Analysis of Strain, Stress and Texture with Quantum Beams, 2nd Edition

Analysis of Strain, Stress and Texture with Quantum Beams, 2nd Edition

Guest Editor

Kenji Suzuki

MDPI

Basel • Beijing • Wuhan • Barcelona • Belgrade • Novi Sad • Cluj • Manchester

Guest Editor
Kenji Suzuki
Faculty of Education
Niigata University
Niigata City
Japan

Editorial Office
MDPI AG
Grosspeteranlage 5
4052 Basel, Switzerland

This is a reprint of the Special Issue, published open access by the journal *Quantum Beam Science* (ISSN 2412-382X), freely accessible at: https://www.mdpi.com/journal/qubs/special_issues/C1O59BLI95.

For citation purposes, cite each article independently as indicated on the article page online and as indicated below:

Lastname, A.A.; Lastname, B.B. Article Title. *Journal Name* **Year**, *Volume Number*, Page Range.

ISBN 978-3-7258-4255-1 (Hbk)
ISBN 978-3-7258-4256-8 (PDF)
https://doi.org/10.3390/books978-3-7258-4256-8

Cover image courtesy of Kenji Suzuki

Contents

About the Editor

Kenji Suzuki

Kenji Suzuki was born in 1958. He graduated from the Department of Mechanical Engineering, Faculty of Engineering, Niigata University in 1980. In 1982, he was hired as an assistant at Niigata University. He was promoted to assistant professor at Niigata University in 1987. He received a Ph.D. from Nagoya University in 1993 for his thesis "X-ray residual stress measurement and strength evaluation of ceramics". He was promoted to professor at Niigata University in 2004. He served as Dean of the Faculty of Education at Niigata University from 2013 to 2017. He retired from Niigata University in 2023 and became a professor emeritus at the same university. He has been studying material strength and X-ray stress measurement, and in the latter half of his career has promoted research into material evaluation using quantum beams. Currently, he is a fellow at Niigata University and a visiting researcher at the Japan Atomic Energy Agency and the Central Research Institute of Electric Power Industry, Japan. He is also the Vice President of the Society for Materials Science, Japan. He has received the Paper Award four times from the Society of Materials Science, Japan, and Paper Award twice from the Japan Society of Maintenology.

Preface

In recent years, there has been remarkable progress in the simulation of material behavior. When we see realistic material deformation and detailed stress maps from simulation results, we feel as if we understand the material behavior. However, the truth can only be known by combining the actual strain and stress with the results of the simulation. History has seen the development of science through the unification of theory and experiment, and the development of materials science is no different. For this reason, the development of experimental research into material deformation behavior and stress analysis is very important. There is always a demand for mutual development between theory and experiment. However, the experimental research is not an even path. Perhaps for this reason, the proportion of experimental research in this field appears to be gradually declining.

When using quantum beams (synchrotron radiation and neutrons), there is a procedure for submitting a proposal, and the proposal may be accepted or rejected. In addition, experiments must be conducted at the beamline, which incurs expenses such as usage fees and travel expenses. Sometimes, when we see these negative aspects, we tend to hesitate to use synchrotron radiation and neutrons. However, quantum beams have excellent usefulness that far exceeds their negative aspects. For example, they have strong penetrating power, high brightness, micro-beams, and the ability to select any wavelength. In addition, the diffraction phenomenon by crystals is based on a clear physical basis, so material behavior can be accurately understood. Quantum beams such as synchrotron radiation and neutrons are excellent means of experimental stress analysis, and their research and industrial applications can contribute to our lives. By using quantum beams, it is now possible to know the stress and strain of coarse grains and weld metals, which were previously difficult to measure.

In Japan, the first SOR-Ring (0.38 GeV) was constructed in 1973 and extracted synchrotron radiations. The energy of the SOR-Ring was low in the ultraviolet and soft X-ray ranges. In 1997, SPring-8 (8 GeV) was constructed. Meanwhile, for neutrons, the research reactor JRR-3 (10 MW) was constructed in 1964, and was upgraded to 20 MW in 1990. J-PARC, the facility that uses spallation neutrons, was constructed, and time-of-flight experiments using neutrons began in 2008. The neutron beam intensity at J-PARC has reached 1 MW. As quantum beam technology improves, detector technology is also advancing rapidly. In the future, we should be able to conduct even better quantum beam experiments.

Following on from the previous Special Issue, "Analysis of Strain, Stress and Texture by Quantum Beams", this current issue includes new Japanese research on materials evaluation using quantum beams. As mentioned above, quantum beam facilities and research using them have been developed in Japan. The materials covered are not only general-purpose metallic materials, but also a wide range of materials such as coarse grains, welding materials, superconducting materials, construction materials, thermal barrier coatings and composite materials. Measurement techniques are based on diffraction and include methodologies such as neutron time-of-flight method, synchrotron radiation double exposure method suitable for coarse grains, extended 3DXRD, and constant penetration depth method.

Analysis of stress and deformation through quantum beam experiments truly proves "seeing is believing". It would bring me much joy if this reprint could be of use to people all over the world.

Kenji Suzuki
Guest Editor

quantum beam science

Editorial

Analysis of Strain, Stress and Texture with Quantum Beams, 2nd Edition

Kenji Suzuki

Factulty of Education, Niigata University, Igarashi-2-no-cho, Nishi-ku, Niigata 950-2181, Japan; suzuki@ed.niigata-u.ac.jp

Welcome to the Special Issue of *Quantum Beam Science*, entitled "Analysis of Strain, Stress and Texture with Quantum Beams, 2nd Edition". The closest international conference to this title is the International Conference on Mechanical Stress Evaluation by Neutron and Synchrotron Radiation (MECASENS). The 9th MECASENS was held from 19 to 21 September 2017 at Skukuza Rest Camp, South Africa [1,2]. The next MECASENS was scheduled to be held in Prague, the Czech Republic, in 2019, but preparations were delayed and it was postponed to 2020. Unfortunately, as you know, the COVID-19 pandemic occurred in 2020. The 10th MECASENS was held on-site and virtually in Prague from 25 to 27 November 2021 [3]. In Japan, domestic and international travel was strictly restricted, and experiments at synchrotron radiation facilities and neutron facilities were also restricted. On the other hand, in 2011, the accident occurred at the Fukushima Daiichi Nuclear Power Plant due to the tsunami caused by the Great East Japan Earthquake, and not only were all nuclear power plants across Japan shut down, but so was the research reactor JRR-3. The JRR-3 finally resumed operation in April 2022.

Due to the above circumstances, experiments using quantum beams to analyze stress, strain, and texture have been restricted, and international exchange has stagnated. This was the most difficult time for quantum beam research. However, even in this difficult situation in Japan, research has continued steadily. This Special Issue can be seen as evidence of that.

Suspension plasma sprayed thermal barrier coatings (SPS-TBCs) have a columnar microstructure, and are therefore expected to have excellent heat cycle and thermal shock resistance [4,5]. However, no investigation has been conducted to evaluate the internal stress distribution in SPS-TBCs with a columnar microstructure. In the Special Issue, the mechanism behind the excellent heat cycle resistance of SPS-TBCs was elucidated using synchrotron radiation and laboratory X-rays [6]. In this experiment, the residual stress was measured using a constant penetration depth method proposed by Ganzel [7,8], and its effectiveness was demonstrated.

A three-dimensional X-ray diffraction (3DXRD) is an excellent X-ray microscope that can measure the shape, orientation, and even strain of crystal grains [9–13]. However, the 3DXRD is limited in the number of crystals and sample dimensions. A noteworthy study has overcome these limitations by incorporating a rotating spiral slit into the scanning 3DXRD technique [14].

Previously, a slit system has been used to create a gauge volume in the bulk material for strain measurement using an area detector [15–17]. As a result, the experiment required a great deal of effort, such as making complex slits and adjustment of the optical system. In this Special Issue, an ingenious method for easily measuring strain in bulk materials is proposed using the area detector without the slit-system [18]. This is the double exposure method (DEM), and it is notable for its applicability to strain measurements in bulk materials, and it has been used to create stress maps of coarse grains and welds.

In crystalline materials with large crystal anisotropy, the stress–strain relationship varies greatly depending on the crystal orientation with plastic deformation. In order to investigate the stress–strain behavior of the lattice planes of a polycrystalline material, it is necessary to measure the elastic–plastic behavior of many diffraction planes. It is well known that intergranular strain is formed as a result of plastic anisotropy [19–21]. Using J-PARC [22], SNS [23], etc., many $(h\,k\,l)$ diffraction peaks can be measured by measuring spallation neutrons with time-of-flight (TOF). Therefore, the TOF is suitable for studying the deformation characteristics of crystalline materials.

Magnesium is the lightest metal in practical use and has the highest specific strength, so it is expected to be used in a wide range of fields. However, it is known that dislocations in magnesium alloys have a strong tendency to slip along their basal planes, resulting in very strong anisotropy of plastic deformation and poor ductility [24,25]. Although studies have been performed on its crystalline elastic–plastic behavior, many aspects remain unclear [26,27]. In HCP-structure magnesium alloys, the increase in lattice strain relative to true stress varies greatly between [*hk.l*] grains; so, in order to obtain true stress, a method has been proposed in which the volume fraction of each grain is weighted using many [*hk.l*] orientations, and the diffraction elastic constant is multiplied. The lattice strain value evaluated from the 12.1 peak shows a good linear relationship with the applied true stress for the whole deformation region [28].

There are also other interesting papers, such as the following, which I highly recommend you read:

1. Nishida, M.; Harjo, S.; Kawasaki, T.; Yamashita, T.; Gong, W. Neutron Stress Measurement of W/Ti Composite in Cryogenic Temperatures Using Time-of-Flight Method. *Quantum Beam Sci.* **2023**, *7*, 8. https://doi.org/10.3390/qubs7010008
2. Yamazaki, Y.; Shinomiya, K.; Okumura, T.; Suzuki, K.; Shobu, T.; Nakamura, Y. Relationship between Internal Stress Distribution and Microstructure in a Suspension-Sprayed Thermal Barrier Coating with a Columnar Structure. *Quantum Beam Sci.* **2023**, *7*, 14. https://doi.org/10.3390/qubs7020014
3. Yasue, A.; Kawakami, M.; Kobayashi, K.; Kim, J.; Miyazu, Y.; Nishio, Y.; Mukai, T.; Morooka, S.; Kanematsu, M. Accuracy of Measuring Rebar Strain in Concrete Using a Diffractometer for Residual Stress Analysis. *Quantum Beam Sci.* **2024**, *8*, 7. https://doi.org/10.3390/qubs7020015
4. Hayashi, Y.; Setoyama, D.; Fukuda, K.; Okuda, K.; Katayama, N.; Kimura H. Scanning Three-Dimensional X-ray Diffraction Microscopy with a Spiral Slit. *Quantum Beam Sci.* **2023**, *7*, 16. https://doi.org/10.3390/qubs7020016
5. Hayashi, Y.; Kimura H. Scanning Three-Dimensional X-ray Diffraction Microscopy for Carbon Steels. *Quantum Beam Sci.* **2023**, *7*, 23. https://doi.org/10.3390/qubs7030023
6. Harjo, S.; Gong. W.; Kawasaki, T. Stress Evaluation Method by Neutron Diffraction for HCP-Structured Magnesium Alloy. *Quantum Beam Sci.* **2023**, *7*, 32. https://doi.org/10.3390/qubs7040032
7. Machiya, S.; Osamura, K.; Hishinuma, Y.; Taniguchi, H.; Harjo, S.; Kawasaki, T. Measurement of Mechanical Behavior of ^{11}B-Enriched MgB_2 Wire Using a Pulsed Neutron Source. *Quantum Beam Sci.* **2023**, *7*, 34. https://doi.org/10.3390/qubs7040034
8. Miura, Y.; Suzuki, K.; Morooka, S.; Shobu, T. Stress Measurement of Stainless Steel Piping Welds by Complementary Use of High-Energy Synchrotron X-rays and Neutrons. *Quantum Beam Sci.* **2024**, *8*, 1. https://doi.org/10.3390/qubs8010001
9. Xu, P.; Zhang, S.; Harjo, S.; Vogel, S.C.; Tomoda, Y. Principal Preferred Orientation Evaluation of Steel Materials Using Time-of-Flight Neutron Diffraction. *Quantum Beam Sci.* **2024**, *8*, 7. https://doi.org/10.3390/qubs8010007

Conflicts of Interest: The author declares no conflicts of interest.

References

1. Marais, D.; Holden, T.M.; Venter, A.M. (Eds.) Mechanical Stress Evaluation by Neutron and Synchrotron Radiation. In *Proceedings of the 9th International Conference on Mechanical Stress Evaluation by Neutron and Synchrotron Radiation, Kruger National Park, South Africa, 19–21 September 2017*; Materials Research Forum LLC: Millersville, PA, USA, 2018; Volume 4, pp. 91–96.
2. MECASENS 2017. Available online: https://na.eventscloud.com/ehome/mecasens/home/ (accessed on 14 December 2024).
3. Abstract Book of MECASENS-10. Available online: https://www.mff.cuni.cz/data//web/obsah/kfm/mecasens10/Book-of-Abstracts-FINAL-Mecasens-2021.pdf (accessed on 14 December 2024).
4. Curry, N.; VanEvery, K.; Snyder, T.; Markocsan, N. Thermal Conductivity Analysis and Lifetime Testing of Suspension Plasma-Sprayed Thermal Barrier Coatings. *Coatings* **2014**, *4*, 630–650. [CrossRef]
5. Gupta, M.; Li, X.-H.; Markocsan, N.; Kjellman, B. Design of high lifetime suspension plasma sprayed thermal barrier coatings. *J. Eur. Ceram. Soc.* **2020**, *40*, 768–779. [CrossRef]
6. Yamazaki, Y.; Shinomiya, K.; Okumura, T.; Suzuki, K.; Shobu, T.; Nakamura, Y. Relationship between Internal Stress Distribution and Microstructure in a Suspension-Sprayed Thermal Barrier Coating with a Columnar Structure. *Quantum Beam Sci.* **2023**, *7*, 14. [CrossRef]
7. Genzel, C. A self-consistent method for X-ray diffraction analysis of multiaxial residual-stress fields in the near-surface region of polycrystalline materials. I. Theoretical concept. *J. Appl. Crystallogr.* **1999**, *32*, 770–778. [CrossRef]
8. Genzel, C.; Broda, M.; Dantz, D.; Reimers, W. A self-consistent method for X-ray diffraction analysis of multiaxial residual-stress fields in the near-surface region of polycrystalline materials. II. Examples. *J. Appl. Crystallogr.* **1999**, *32*, 779–787. [CrossRef]
9. Poulsen, H.F.; Nielsen, S.F.; Lauridsen, E.M.; Schmidt, S.; Suter, R.M.; Lienert, U.; Margulies, L.; Lorentzen, T.; Jensen, D.J. Three-dimensional maps of grain boundaries and the stress state of individual grains in polycrystals and powders. *J. Appl. Crystallogr.* **2001**, *34*, 751–756. [CrossRef]
10. Alpers, A.; Knudsen, E.; Poulsen, H.F.; Herman, G.T. Resolving ambiguities in reconstructed grain maps using discrete tomography. *Electron. Notes Discret. Math.* **2005**, *20*, 419–437. [CrossRef]
11. Poulsen, H.F. *Three-Dimensional X-Ray Diffraction Microscopy, Mapping Ploycrystals and Their Dynamics*; Springer: Berlin/Heidelberg, Germany, 2004.
12. Poulsen, H.F.; Ludwig, W.; Schmidt, S. 3D X-ray diffraction microscope. In *Neutron and Synchrotron Radiation in Engineering Materials Science*; Reimers, W., Pyzalla, A.R., Schereyer, A., Clemens, H., Eds.; Wiley-VCH Verlag GmbH & Co. KGaA: Weinheim, Germany, 2008; pp. 335–351.
13. Poulsen, H.F.; Ludwig, W.; Lauridsen, E.M.; Schmidt, S.; Pantleon, W.; Olsen, U.L.; Oddershede, J.; Reischig, P.; Lyckegaard, A.; Wright, J. 4D characterization of metals by 3DXRD. *Proc. 31st RisØ Int. Symp. Mater. Sci.* **2010**, *31*, 101–119.
14. Hayashi, Y.; Setoyama, D.; Fukuda, K.; Okuda, K.; Katayama, N.; Kimura, H. Scanning Three-Dimensional X-ray Diffraction Microscopy with a Spiral Slit. *Quantum Beam Sci.* **2023**, *7*, 16. [CrossRef]
15. Suzuki, K.; Shobu, T.; Shiro, A.; Toyokawa, H. Evaluation of Internal Stresses Using Rotating-Slit and 2D Detector. *Mater. Sci. Forum* **2013**, *772*, 15–19. [CrossRef]
16. Martines, R.V.; Honkimäki, V. Depth resolved strain and phase mapping of dissimilar friction stir welds using high energy synchrotron radiation. *Texture Microstruct.* **2003**, *35-3/4*, 145–152. [CrossRef]
17. Martins, R.V.; Ohms, C.; Decroos, K. Full 3D spatially resolved mapping of residual strain in a 316L austenitic stainless steel weld specimen. *Mater. Sci. Eng.* **2010**, *527*, 4779–4787 [CrossRef]
18. Miura, Y.; Suzuki, K.; Morooka, S.; Shobu, T. Stress Measurement of Stainless Steel Piping Welds by Complementary Use of High-Energy Synchrotron X-rays and Neutrons. *Quantum Beam Sci.* **2024**, *8*, 1. [CrossRef]
19. Corvasce, F.; Lipinsky, P.; Berveiller, M. Intergranular residual stresses in plastically deformed polycrystals. In *Residual Stresses II*; Beck, G., Denis, S., Simon, A., Eds.; Elsevier: Amsterdam, The Netherlands, 1989; pp. 535–541.
20. Holden, T.M. Intergranular stress. *J. Neutron Res.* **1999**, *7*, 291–317. [CrossRef]
21. Hounkpati, V.; Fréour, S.; Gloaguen, D.; Legrand, V.; Kelleher, J.; Kockelman, W.; Kabra, S. In situ neutron measurements and modelling of the intergralular strains in the near-β titanium alloy Ti-β21S. *Acta Mater.* **2016**, *109*, 341–352. [CrossRef]
22. J-PARC. Available online: https://j-parc.jp/c/en/index.html (accessed on 14 December 2024).
23. SNS. Available online: https://neutrons.ornl.gov/sns (accessed on 14 December 2024).
24. Agnew, S.R.; Duygulu, Ö. Plastic anisotropy and the role of non-basal slip in magnesium alloy AZ31B. *Int. J. Plast.* **2005**, *21*, 1161–1193. [CrossRef]
25. Sandlöbes, S.; Friák, M.; Neugebauer, J.; Paabe, D. Basal and non-basal dislocation slip in Mg-Y. *Mater. Sci. Eng.* **2013**, *576*, 61–68. [CrossRef]
26. Brown, D.W.; Agnew, S.R.; Bourke, M.A.M.; Holden, T.M.; Vogel, S.C.; Tmomé, C.N. Internal strain and texture evolution during deformation twinning in magnesium. *Mater. Sci. Eng.* **2005**, *399*, 1–12. [CrossRef]

27. Woo, W.; Choo, H.; Prime, M.B.; Feng, Z.; Clausen, B. Microstructure, texture and residual stress in a friction-stir-processed AZ31B magnesium alloy. *Acta Mater.* **2008**, *56*, 170–1711. [CrossRef]
28. Harjo, S.; Gong, W.; Kawasaki, T. Stress Evaluation Method by Neutron Diffraction for HCP-Structured Magnesium Alloy. *Quantum Beam Sci.* **2023**, *7*, 32. [CrossRef]

quantum beam science

MDPI

Article

Scanning Three-Dimensional X-ray Diffraction Microscopy for Carbon Steels

Yujiro Hayashi † and Hidehiko Kimura *

Toyota Central R&D Labs., Inc., 41-1, Yokomichi, Nagakute 480-1192, Japan; y-hayashi@spring8.or.jp
* Correspondence: hdkimura@mosk.tytlabs.co.jp
† Current address: RIKEN SPring-8 Center, 1-1-1 Kouto, Sayo-cho, Sayo-gun 679-5148, Japan.

Abstract: Plastically deformed low-carbon steel has been analyzed by nondestructive three-dimensional orientation and strain mapping using scanning three-dimensional X-ray diffraction microscopy (S3DXRD). However, the application of S3DXRD is limited to single-phase alloys. In this study, we propose a modified S3DXRD analysis for dual-phase alloys, such as ferrite–pearlite carbon steel, which is composed of grains detectable as diffraction spots and a phase undetectable as diffraction spots. We performed validation experiments for ferrite–pearlite carbon steel with different pearlite fractions, in which the ferrite grains and the pearlite corresponded to the detectable grains and an undetectable phase, respectively. The regions of pearlite appeared more remarkably in orientation maps of the ferrite grains obtained from the carbon steel samples than that of the single-phase low-carbon steel and increased with the increase in the carbon concentration. The fractions of the detectable grains and the undetectable phase were determined with an uncertainty of 15%–20%. These results indicate that the proposed modified analysis is qualitatively valid for dual-phase alloys comprising detectable grains and an undetectable phase.

Keywords: 3DXRD; scanning 3DXRD; orientation; carbon steel; dual phase

1. Introduction

Predicting damage and failure in polycrystalline alloys, which are widely used in various industries, is challenging due to not only the difficulty in modeling the dynamics of numerous dislocations in hierarchical microstructures with a length scale ranging from nanometers to millimeters or more but also the lack of suitable geometric models of three-dimensional (3D) microstructures and validation data for the mechanical responses. Three-dimensional polycrystalline microstructures can be reconstructed by fully automated serial sectioning [1,2] with the aid of 3D orientation microscopy (OM) based on electron backscatter diffraction (EBSD) [3]. The distribution of geometrically necessary dislocations can also be determined by postprocessing the EBSD OM data. In addition, high-angular resolution EBSD (HR-EBSD) can detect local elastic strains [4]. However, mechanical responses, such as plastic deformation, creep, fatigue, and fracture, cannot be tracked by destructive serial-sectioning microscopy.

In the past two decades, nondestructive 3D OM and strain mapping for polycrystalline alloys have been achieved using synchrotron-based X-ray diffraction (XRD) and imaging methods, such as differential aperture X-ray microscopy (DAXM) [5–13], diffraction contrast tomography (DCT) [14–20], 3D XRD microscopy (3DXRD) [21–43], high-energy diffraction microscopy (HEDM) [44–66], and scanning 3DXRD [67–77]. DAXM uses a focused polychromatic X-ray beam and a far-field area detector to obtain Laue diffraction patterns from multiple grains. Three-dimensional OM can be performed without serial sectioning by scanning a platinum wire near the sample surface and applying indexing for multiple grains (multigrain indexing) to the Laue diffraction patterns. Although, in DAXM, sample rotation is unnecessary, and the spatial resolution is determined by the incident beam size, nondestructive OM is limited to subsurface grains because polychromatic X-ray illumination of numerous grains causes a considerable overlap of diffraction spots on the detector.

In DCT, 3DXRD, HEDM, and scanning 3DXRD with high-energy monochromatic X-ray beams, 3D OM is performed by detecting diffracted beams from multiple grains deeply embedded in a sample. These XRD and imaging methods differ in terms of the shape of the incident X-ray beam and the spatial resolution of the detector. In DCT, a box beam illuminates a polycrystalline sample, and diffracted beams from grains are detected using a near-field area detector with high spatial resolution. A 3D grain map can be obtained by applying a forward projection model to the diffraction patterns obtained from a single rotation scan. Although, in DCT, a 3D grain map can be obtained from a single rotation scan, its application is limited to lightly deformed samples with slight intragranular misorientation. This is because it is difficult to achieve a 3D reconstruction for heavily-deformed samples owing to the overlap of diffraction spots and the peak broadening caused by intragranular misorientation.

In 3DXRD and HEDM, a line-focused beam illuminates a sample, and the diffraction patterns are obtained by a layer-by-layer scan (with sample rotations) using near- and far-field area detectors. A 3D orientation map can be obtained from 3DXRD scan data by applying layer-by-layer multigrain indexing on diffraction spots from the far-field detector and then layer-by-layer reconstruction of the grain positions and shapes with or without intragranular misorientation by applying ray tracing, an algebraic reconstruction technique, or forward modeling to the diffraction patterns from the near-field detector. Compared with DCT, 3DXRD can obtain orientation maps with intragranular misorientation caused by plastic deformation. Thus, 3DXRD and HEDM have enabled 3D orientation mapping with mechanical responses such as slip and twin deformation, shock deformation, creep, fatigue, and fracture. However, 3DXRD and HEDM have been performed on small samples such as matchstick-like samples, wire materials, or miniature test pieces to suppress the overlap of broadened diffraction spots.

In scanning 3DXRD, a point-focused (or pencil) beam is irradiated on a sample, far-filed diffraction patterns are obtained from 3D scans with sample rotations and translations, and a 3D orientation map with intragranular misorientation can be obtained by applying voxel-by-voxel multigrain indexing to the scanning 3DXRD data. In scanning 3DXRD, the number of illuminated grains is reduced using the point-focused beam, thus suppressing the overlap of diffraction spots. In addition, the measurement using a far-field detector can be combined with a conical slit [78]. The conical slit can considerably reduce the overlap of the diffraction spots by shielding the detector from the diffracted beams of many grains except for those from grains inside the region of interest called the gauge volume. Therefore, scanning 3DXRD with a conical slit is most suitable for samples with numerous strained grains at the expense of the scanning time.

One of next key steps of scanning 3DXRD is its application to dual-phase alloys. Ferrite–pearlite carbon steel hardenable by quenching has wide applications due to its low cost, and the ferrite–martensite dual-phase steel is used in the automotive industry owing to its balance between ductility and strength. In most XRD OM experiments, diffraction spots from nondeformed ferrite grains are detectable, and boundaries between the ferrite grains can be reconstructed. Therefore, nondestructive 3D orientation and strain mapping of single-phase low-carbon steel consisting of ferrite grains have been achieved [73]. Hereafter, we refer to ferrite grains as *visible* grains. Pearlite and martensite correspond to *invisible* phases because the diffraction spots cannot be detected from pearlite and martensite in XRD OM. In nondestructive tomographic XRD OM techniques (DCT, 3DXRD, HEDM, and scanning 3DXRD), *invisible* phases cause reconstruction errors, *visible* grains may abnormally grow into the regions of *invisible* phases, and in the worst case, the regions of *invisible* phases can disappear.

In this study, we propose a modified scanning 3DXRD analysis that can be applied to dual-phase alloys containing an *invisible* phase, such as ferrite–pearlite carbon steel.

In this analysis, we introduced the threshold of completeness as a reconstruction parameter, where completeness represents the probability of the existence of *visible* grains. Herein, we performed qualitative validation using ferrite–pearlite carbon steel with different pearlite fractions by comparing the fractions of *visible* grains in the obtained orientation maps.

2. Materials and Methods

2.1. Materials

A ferrite single-phase low-carbon steel sample was obtained from a commercial cold-rolled steel sheet defined by Japanese industrial standards (JIS) G3141. A uniaxial sheet tensile-test piece with a width of 1 mm at the gauge section was cut from a 1 mm thick cold-rolled steel sheet using a wire cutter. The sheet tensile-test piece was used as a ferrite single-phase low-carbon steel sample because the fine spheroidal cementite particles in ferrite grains can be neglected.

Ferrite–pearlite carbon steel samples were obtained from commercial round bar shaped carbon steel defined as S15C, S25C, S35C, and S45C by JIS G4051 to apply the proposed analysis to commercially available alloys. The carbon contents of S15C, S25C, S35C, and S45C are defined as 0.13–0.18, 0.22–0.28, 0.32–0.38, and 0.42–0.48 wt%, respectively. The carbon contents indicate that the pearlite fraction of S45C is higher than that of S35C, as shown in Figure 1. From the carbon contents defined by JIS G4051, the pearlite fraction of S35C is higher than that of S25C, which is higher than that of S15C. The round bars were cut into uniaxial round-bar tensile-test pieces with a 1 mm diameter at the gauge section using a wire cutter. The S15C, S25C, S35C, and S45C samples were used as ferrite–pearlite carbon steel samples with different pearlite fractions.

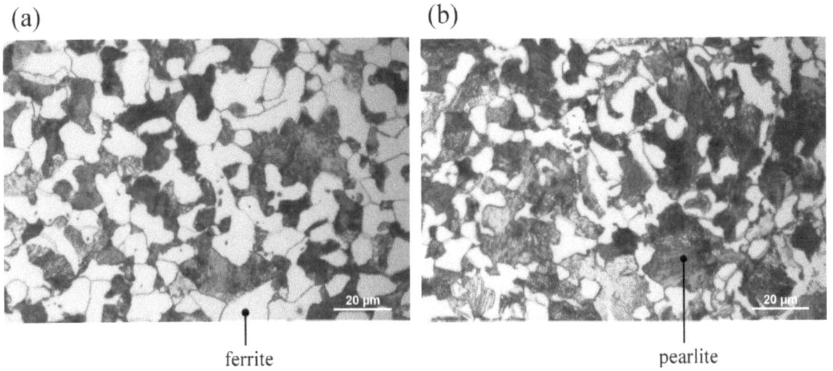

Figure 1. Optical images of the (**a**) S35C and (**b**) S45C samples. The white and black regions indicate ferrite grains and pearlite, respectively. The fractions of ferrite grains in the S35C and S45C samples were 55% and 33%, respectively.

2.2. Data Acquisition

We conducted experiments using a scanning 3DXRD apparatus equipped with X-ray-focusing mirrors at BL33XU of SPring-8. Radiation from an in-vacuum undulator was monochromatized with a Si(311) double crystal monochromator to a monochromatic X-ray beam with 50 keV of photon energy. The monochromatic beam was focused with 400 mm long Pt-coated Kirkpatrick-Baez mirrors into a beam size of $\sim 1 \times 1\ \mu m^2$, as shown in Figure 2. The X-ray microbeam was irradiated on the gauge section of the sample mounted on the rotation stage with the longitudinal direction of the sample parallel to the z axis and the incoming microbeam parallel to the y axis (Figure 3). A conical slit [78] was placed 50 mm away from the sample to prevent the detection of diffracted beams from grains outside the gauge volume. The aperture of the conical slit was 30–110 μm, depending on the Bragg angles. The Bragg angles of the conical slit were 3.506°, 4.958°, 6.074°, 7.011°, 7.844°, 8.594°, and 9.284° for the 110, 200, 211, 220, 310, 222, and 321 reflections of α-Fe and

50 keV. The designed lengths of the gauge volume in the y axis were 330–360 μm. An X-ray flat-panel detector (2923NDT, Dexela) with an active image area of 291 × 230 mm^2 and a pixel size of 150 × 150 μm^2 at a 2 × 2 binning mode was placed downstream of the conical slit as the far-field detector. The sample was rotated at a constant speed along the z axis. The rotation speed was $16°s^{-1}$ for the S45C sample and $20°s^{-1}$ for the other samples. Images from the detector were recorded every 0.6° during the sample rotation of ω = 0°–180°. The exposure time per image was 38 and 30 ms for the S45C and the other samples, respectively. The diffracted beams that passed through the conical slit were recorded as diffraction spots on the images. When ω reached 180°, the sample was translated by 1 μm in the x direction using the translation stage located below the ω rotation stage. This procedure was repeated for an x range of $-x_m \le x \le x_m$, where x_m is 150 μm for the low-carbon steel sample and 100 μm for the carbon steel samples. Thus, diffraction images containing diffraction spots from ferrite grains were recorded as a function of ω and x. From the two-dimensional (2D) ω–x scan data, a 2D orientation map with a voxel size of 1 μm and a field of view (FoV) of $2x_m$ in diameter in the xy plane was obtained. On the orientation map, ferrite grains and pearlite correspond to the *visible* grains and the *invisible* phase, respectively (Optionally, a 3D orientation map can be obtained by stacking the 2D orientation maps.)

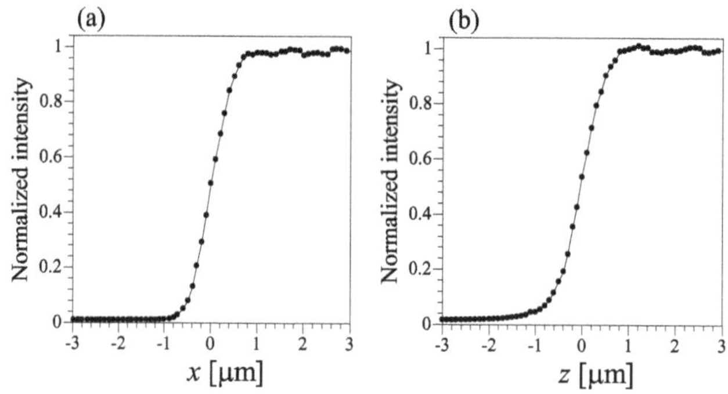

Figure 2. (**a**) Horizontal and (**b**) vertical knife-edge scan profiles of a focused X-ray beam with 50 keV of photon energy.

2.3. Concept of Modified Completeness

The modified completeness is based on the completeness $N/M = N'$ of the conventional scanning 3DXRD [69], where N is the number of detected diffraction spots for a *visible* grain, and M is the theoretically expected maximum of N. The completeness represents the probability of the existence of *visible* grains. Generally, the completeness curves of individual grains are continuously distributed from the inside to the outside of the grains. The grain boundaries between the *visible* grains appear where neighboring completeness curves cross each other. In ideal cases, without experimental errors, N' is almost 1 inside the *visible* grains and sharply drops outside the grains (Figure 4a). Then, the grain boundaries can be determined even for a sample containing an *invisible* phase (Figure 4b). In real cases, however, N' may not be up to 1 and gradually decreases from the center of grains to the outside of the grains owing to the insufficient diffraction intensities, the insufficient quantum efficiency, and the noise of the detector, as well as the short exposure time (Figure 4c). These experimental limitations result in orientation maps with reconstruction errors or artifacts, in which small grains close to large grains shrink or disappear, large grains close to small grains grow, and the concave and convex shapes of grains disappear. As artifacts exist more in samples containing an *invisible* phase, we propose $N'/N'_{max}(i)$ as the modified completeness, where the denominator $N'_{max}(i)$ is the maximum of N' of each grain, and i indicates that N'_{max} depends on the i-th grain. Using the modified completeness, the artifacts can be reduced (Figure 4d). Thus, we introduce the threshold of completeness N'_{th}, which allows the reconstruction of grain boundaries between the *visible* grains and the *invisible* phase (Figure 4e). In real cases with experimental errors, the completeness threshold may not be determined for all the *visible* grains in the FoV, resulting in uncertainties in the completeness threshold and the positions of the grain boundaries between the *visible* grains (Figure 4f), as well as the boundaries between the *visible* grains and the *invisible* phase (Figure 4g).

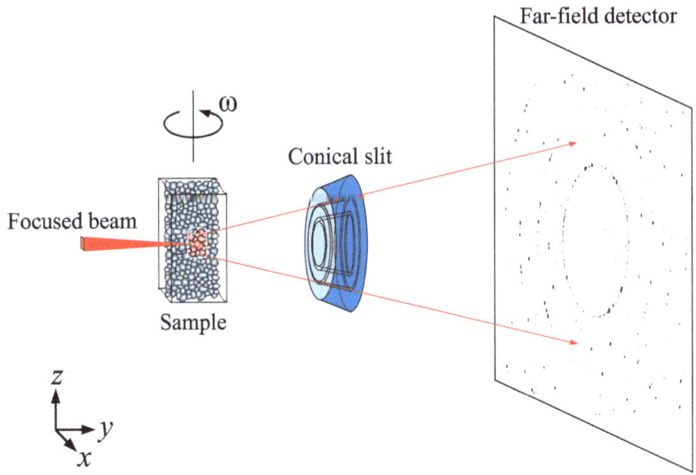

Figure 3. Schematic of the experimental setup. The focused X-ray microbeam parallel to the y axis is irradiated on a polycrystalline sample rotated around the z axis. The sample rotation angle is defined as ω. Diffracted beams from multiple ferrite grains in the sample pass through the conical slit for α-Fe and are detected as diffraction spots in the far-field detector. In this scanning 3DXRD setup, ferrite grains and pearlite correspond to the *visible* grains and the *invisible* phase, respectively.

2.4. Data Analysis

From the obtained ω–x scan data, the conventional completeness $N/M = N'$ was evaluated for each voxel and the *visible* grains using the conventional scanning 3DXRD method [69]. N and M were calculated with *ImageD*11 [79] and *PolyXSim* [80] software, respectively. Thereafter, each *visible* grain was extracted using the map of N' as follows. First, a voxel was randomly selected from the map. Among the candidate grains at the selected voxel, the grain with the highest N' was selected. Next, the misorientation between the grain at the selected voxel and those at the first neighboring voxels was calculated. If the misorientation was less than 2° as intragranular misorientation, the grain at the selected voxel and those at the first neighboring voxels were considered the same. This was also applied to the misorientation between the first and second neighboring voxels. The above steps were repeated until there was no new neighboring voxel. If the number of voxels regarded as voxels in the same grain was greater than the threshold of a grain size, the voxels were extracted as a single grain existing in the xy plane illuminated by the focused X-ray microbeam because a large neighboring grain in the z direction may appear as a minute grain due to the tail of the microbeam in the z direction (Figure 2b). Then, the extracted grain was deleted from the original map. The normalization factor N'_{max} was evaluated for the extracted single grain, and the distribution of completeness $N'(x, y)/N'_{max}$ for the grain was determined. Next, a voxel was randomly selected from the map from which a single grain was deleted, and then another single grain was extracted. The extractions of single grains were repeated until no new grains were found. Thus, an orientation map with information about the completeness of each grain was obtained.

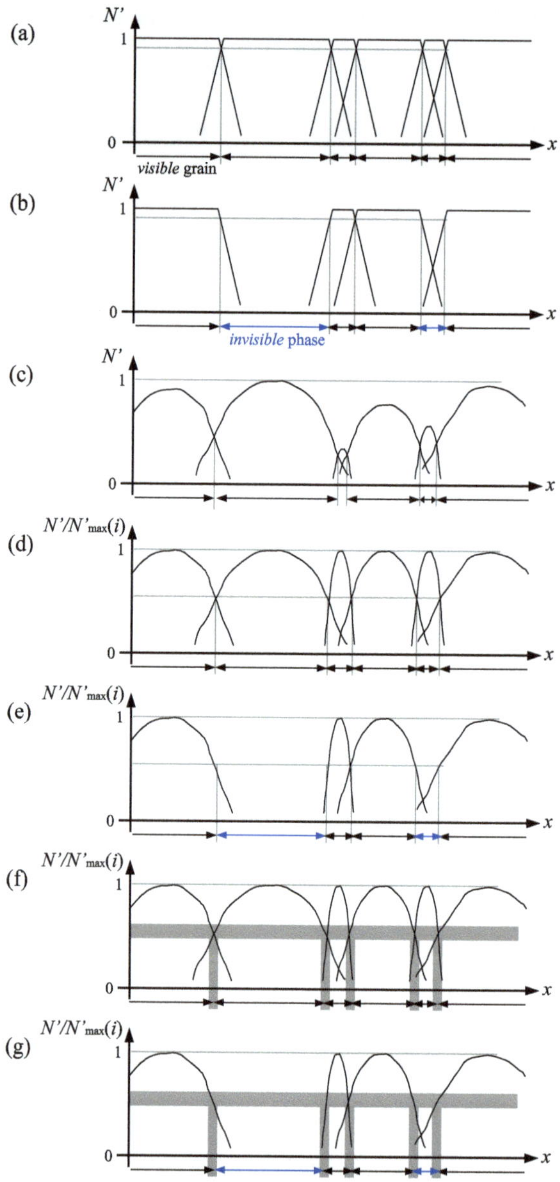

Figure 4. Schematic of the completeness curves of *visible* grains for ideal cases (**a**) without and (**b**) with an *invisible* phase, (**c**) a real case without the *invisible* phase using conventional completeness N', expected cases (**d**) without and (**e**) with an *invisible* phase using the modified completeness $N'/N'_{\max}(i)$ and its threshold, and real cases (**f**) without and (**g**) with the *invisible* phase owing to experimental errors. In the cases of (**a**,**b**), the completeness of the *visible* grains represents almost 1, and the boundaries between the *visible* grains and the *invisible* phase are observed. In the case of (**c**), the completeness depends on the grain size. In the cases of (**d**,**e**), the maximum completeness of each grain is 1, and the boundaries between the *visible* grains and the *invisible* phase can be determined using the threshold of completeness. In the case of (**f**,**g**), there are nonnegligible uncertainties for the threshold and the positions of the boundaries.

3. Results

Figure 5 shows the orientation maps of the single-phase low-carbon steel sample with different completeness thresholds (N'_{th}). The orientation is represented by the z direction in the basic triangle of the pole figure of α-Fe. The colored pixels with orientation information and the uncolored white pixels indicate the regions above and below the completeness thresholds, respectively. On the maps with N'_{th} = 0.98–0.99, the centers of the *visible* grains appeared, but the grain boundaries could not be observed. On the maps with N'_{th} = 0.94–0.96, the grain boundaries were visible, except for the grain-boundary triple points. On the maps with N'_{th} = 0.89–0.92, the grain-boundary triple points were visible. These results showed that the optimal completeness threshold for most grain boundaries ranged from 0.90 to 0.95, and there was an uncertainty of about 0.05 for the optimal completeness threshold. On the map with N'_{th} = 0.85, the grain-boundary triple points were clear, and the fraction of *visible* grains was more than 99% (Figure 5g). On the map with N'_{th} = 0.85, the ferrite grains are indicated by regions enclosed by the same orientation (or the same color), and the grain boundaries are indicated by the boundaries between the different orientations (or different colors). Polygonal grains and clear grain boundaries, which are common for annealed single-phase polycrystalline metals and alloys, were observed. The regions of less than 1% at some grain boundaries and particularly the grain-boundary triple points are indicated by the uncolored regions below the threshold, suggesting reconstruction errors rather than an *invisible* phase owing to insufficient diffraction intensities.

Figures 6–9 show orientation maps with completeness thresholds obtained from the S15C, S25C, S35C, and S45C carbon steel samples, respectively. The samples showed a similar relationship between the formation of the grain boundaries and N'_{th}, except for the fractions of the *visible* grains. For the same threshold, the fraction of the *visible* grains decreased in the order of S15C, S25C, S35C, and S45C. At N'_{th} = 0.85, the fractions of the *visible* grains were 94%, 91%, 82%, and 58% for the S15C, S25C, S35C, and S45C samples, respectively, which were much smaller than that of the low-carbon steel sample. Therefore, the uncolored regions below N'_{th} = 0.85 in the carbon steel samples were attributed to pearlite.

Figure 10 shows the variation in the fraction of the *visible* grains with the completeness threshold. In the N'_{th} range of 0.85–0.95, the fraction of the *visible* grains decreased with the carbon concentration in the order of S15C, S25C, S35C, and S45C. This result qualitatively validates the concept of modified completeness and its threshold. The uncertainty of 0.05 for the optimal completeness threshold corresponded to an uncertainty of 15%–20% for the fraction of *visible* grains at an optimal N'_{th} range of 0.90–0.95. The fractions of the ferrite grains determined from the optical photographs (55% and 33% for the S35C and S45C carbon steel samples) corresponded to an N'_{th} of 0.93–0.94. The fraction of the *visible* grains in the low-carbon steel sample exceeded 85% at N'_{th} = 0.93–0.94. These results were consistent with the uncertainties in the fractions of the *visible* grains and the optimal completeness threshold.

When N'_{th} was reduced from 0.99 to 0.94, the fraction of the *visible* grains in the low-carbon steel sample increased sharply, reaching a plateau in the N'_{th} range of 0.89–0.94. This saturation point of the fraction curve at N'_{th} = 0.94 corresponded to N'_{th}, at which most of the *visible* grain boundaries were formed, except for the grain-boundary triple points. Although the saturation point became unclear in the order of S15C, S25C, S35C, and S45C samples, it was obvious at the same threshold of N'_{th} = 0.94 for the low-carbon steel sample and the S15C and S25C carbon steel samples.

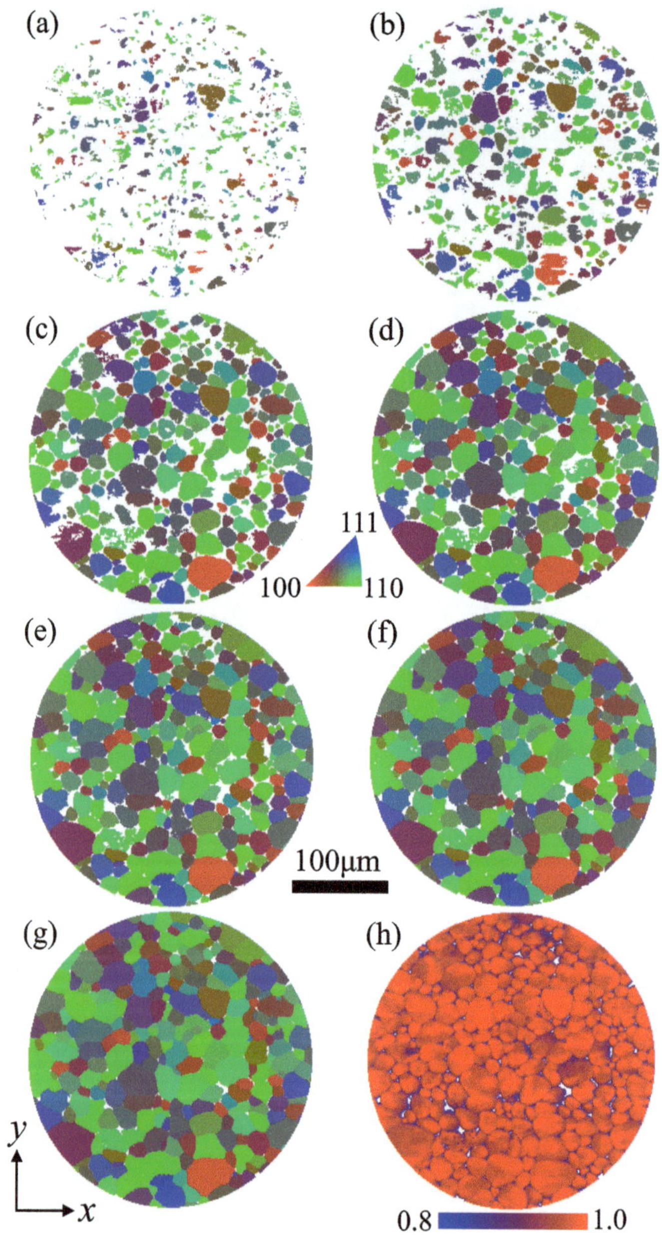

Figure 5. Orientation maps obtained from the single-phase low-carbon steel sample with completeness thresholds N'_{th} = (**a**) 0.99, (**b**) 0.98, (**c**) 0.96, (**d**) 0.94, (**e**) 0.92, (**f**) 0.89, and (**g**) 0.85 and (**h**) the completeness map. The FoV diameter and pixel size are 300 µm and 1 × 1 µm^2, respectively. The orientation of the *visible* grains is represented by the z axis in the basic triangle of the pole figure of α-Fe. The uncolored white pixels indicate the regions below the completeness thresholds.

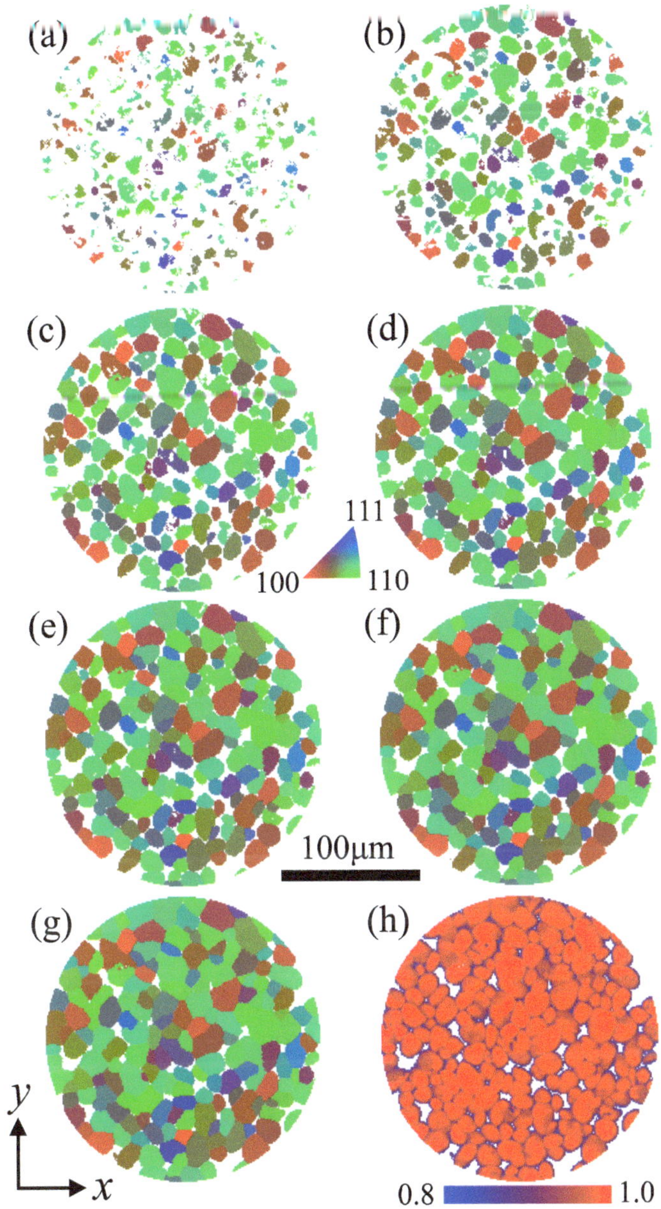

Figure 6. Orientation maps obtained from the S15C carbon steel sample with completeness thresholds N'_{th} of (**a**) 0.99, (**b**) 0.98, (**c**) 0.96, (**d**) 0.94, (**e**) 0.92, (**f**) 0.89, and (**g**) 0.85 and (**h**) the completeness map. The FoV diameter and pixel size are 200 μm and 1 × 1 μm^2, respectively. The orientations of the *visible* grains are represented by the z axis in the basic triangle of the pole figure of α-Fe, and the uncolored white pixels indicate the regions below the completeness thresholds.

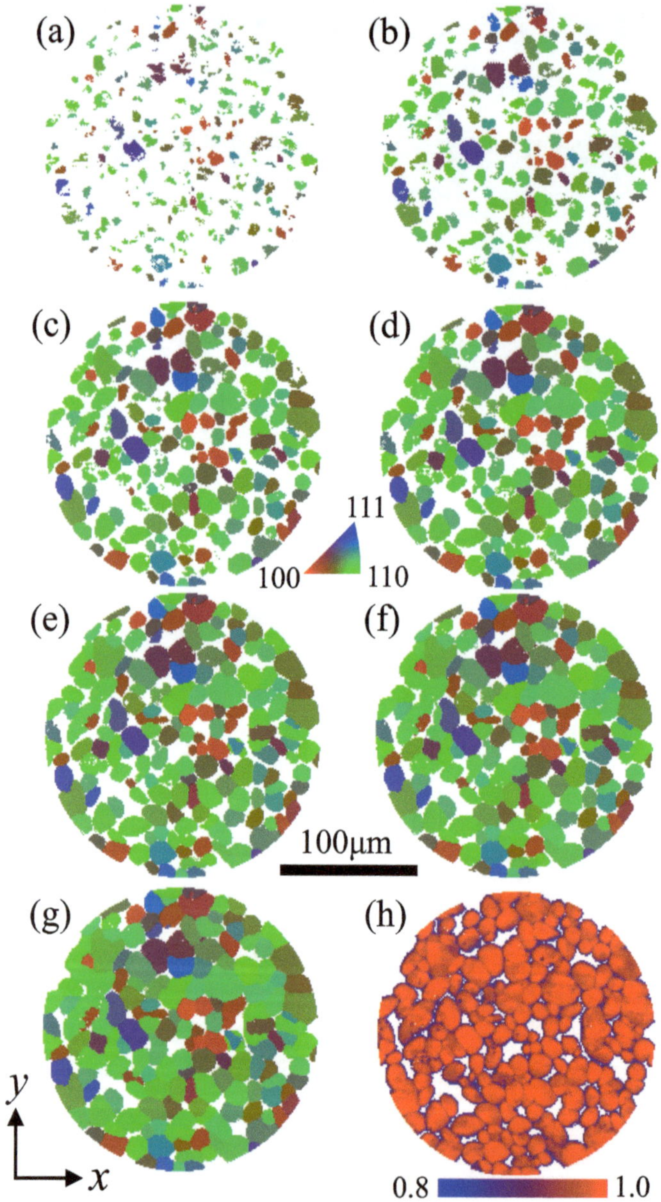

Figure 7. Orientation maps obtained from the S25C carbon steel sample with completeness thresholds N'_{th} of (**a**) 0.99, (**b**) 0.98, (**c**) 0.96, (**d**) 0.94, (**e**) 0.92, (**f**) 0.89, and (**g**) 0.85 and (**h**) the completeness map. The FoV diameter and pixel size are 200 μm and 1×1 μm^2, respectively. The orientations of the *visible* grains are represented by the z axis in the basic triangle of the pole figure of α-Fe, and the uncolored white pixels indicate the regions below the completeness thresholds.

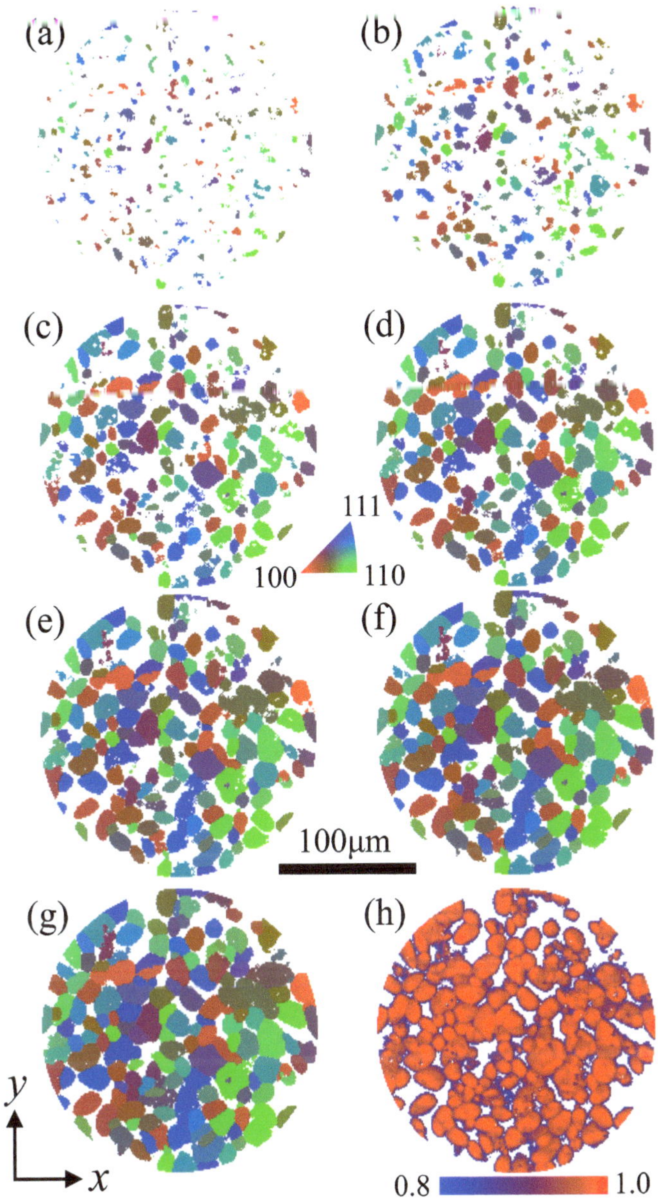

Figure 8. Orientation maps obtained from the S35C carbon steel sample with completeness thresholds N'_{th} of (**a**) 0.99, (**b**) 0.98, (**c**) 0.96, (**d**) 0.94, (**e**) 0.92, (**f**) 0.89, and (**g**) 0.85 and (**h**) the completeness map. The FoV diameter and pixel size are 200 μm and 1×1 μm^2, respectively. The orientations of the *visible* grains are represented by the z axis in the basic triangle of the pole figure of α-Fe, and the uncolored white pixels indicate the regions below the completeness thresholds.

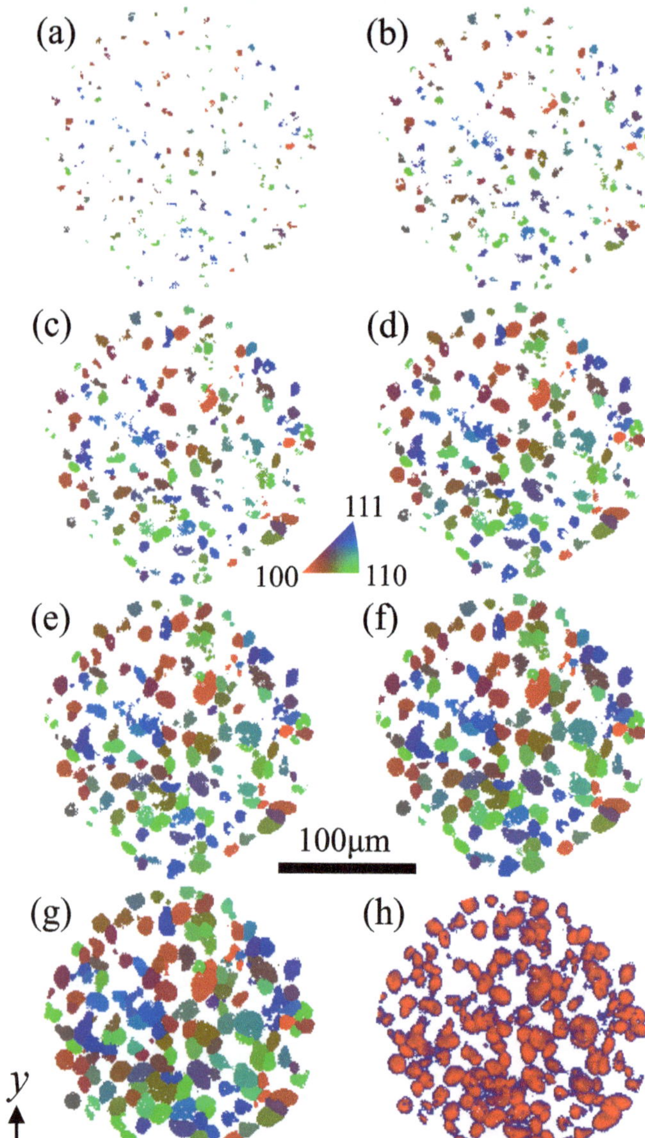

Figure 9. Orientation maps obtained from the S45C carbon steel sample with completeness thresholds N'_{th} of (**a**) 0.99, (**b**) 0.98, (**c**) 0.96, (**d**) 0.94, (**e**) 0.92, (B) 0.89, and (**g**) 0.85 and (**h**) the completeness map. The FoV diameter and pixel size are 200 μm and 1 × 1 μm^2, respectively. The orientations of the *visible* grains are represented by the *z* axis in the basic triangle of the pole figure of α-Fe, and the uncolored white pixels indicate the regions below the completeness thresholds.

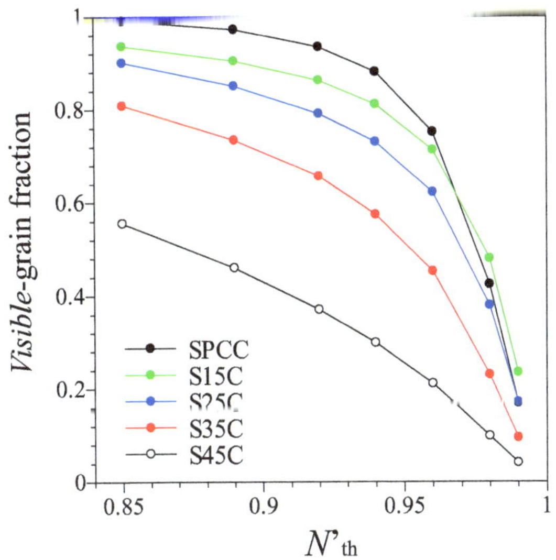

Figure 10. Dependence of the fraction of the *visible* grains on the completeness threshold for the single-phase low-carbon steel sample and the S15C, S25C, S35C, and S45C carbon steel samples.

4. Discussion

The concept of modified completeness and its threshold were quantitatively validated. However, there were two major discrepancies between the expected and obtained results: (1) the fractions and (2) the shapes of the *visible* grains. The reconstruction of the former was limited due to the uncertainty of 0.05 for the completeness threshold, which corresponds to the uncertainty of 15–20% for the fractions of the *visible* grains. This is the main limitation of using a simple definition of completeness and a single threshold value for all grain boundaries. For more quantitative reconstruction, there is a need for more sophisticated methods, such as forward modeling, which demands much higher computational costs compared with the proposed analysis method based on multigrain indexing using a simple definition of completeness and a single threshold value for all grain boundaries.

In the optimal N'_{th} range of 0.90–0.95, the fraction of the *visible* grains was 80–90% for the single-phase low-carbon steel sample. The uncolored regions below $N'_{th} = 0.95$ were mostly located at the grain-boundary triple points, indicating that the grain shapes at the optimal threshold were limited to round shapes compared with the polygonal shapes of the *visible* grains. The discrepancy in the grain shapes was also observed in the optical photograph and the obtained orientation map for the S45C carbon steel sample. The reproducibility of the shapes of the *visible* grains follows the performance of conventional scanning 3DXRD, in which concave and convex grain shapes disappear even for the single-phase low-carbon sample. These artifacts yield round *visible* grains.

The artifacts in the shapes of the *visible* grains can also be attributed to experimental errors. In ideal cases, without experimental errors, N' is 1 in the *visible* grains and drops sharply outside the grains. In real cases, insufficient diffraction intensities result in blurred grain boundaries, which cause coarsening (or shrinking) of grains and more uncertainty in the optimal completeness threshold.

We infer that these limitations caused an uncertainty of 0.05 for the optimal threshold and 15–20% for the fractions of the *visible* grains. Considering the uncertainty, the optimal threshold can be defined by the threshold at the saturation point of the *visible*-grain fraction curve at the expense of the polygonal grain shapes because most *visible* grain boundaries are formed at the saturation point, except for the grain-boundary triple points. Although the saturation point was unclear for the carbon steel samples with large pearlite fractions, it was obvious for the carbon steel samples with small pearlite fractions. Thus, the optimal threshold can be determined using not only single-phase alloys but also alloys with small fractions of *invisible* phases as reference samples even for alloys with large fractions of *invisible* phases.

5. Conclusions

We propose a modified scanning 3DXRD as a nondestructive OM method applicable to dual-phase alloys, such as ferrite–pearlite carbon steel, in which ferrite grains and pearlite correspond to the *visible* grains and an *invisible* phase because the diffraction spots from ferrite grains are detectable, whereas those from the pearlite are not. We defined $N'/N'_{\text{max}}(i)$ as the modified completeness and introduced its threshold N'_{th} for the *visible* grains, where the number of detected diffraction spots from each *visible* grain, N', was normalized by the maximum value of N' of the i-th grain ($N'_{\text{max}}(i)$). In the proposed modified analysis, N'_{th} was determined as the completeness at the boundaries between the *visible* grains and applied to all the *visible* grains close to the *invisible* phase, which allowed the reconstruction of the boundaries between the *visible* grains and the *invisible* phase. The concept of modified completeness and its threshold was validated with an uncertainty of 15%–20% for the fraction of *visible* grains from the orientation maps of ferrite single-phase and ferrite–pearlite carbon steel with different pearlite fractions. The main limitations for more quantitative reconstruction are attributed to the uncertainty for the determination of N'_{th} for all the *visible* grains and the artifacts resulting in round *visible* grains.

Author Contributions: Y.H. designed the research and performed the experiments and analysis. H.K. managed the research. All authors discussed the results and wrote the manuscript. All authors have read and agreed to the published version of the manuscript.

Funding: This research received no external funding.

Data Availability Statement: The data presented in this manuscript are available on request from the corresponding author. The data are not publicly available due to the policy of the authors' affiliation (an incorporated company).

Acknowledgments: Synchrotron radiation experiments were performed at the BL33XU beamline of SPring-8 with the approval of the Japan Synchrotron Radiation Institute (Proposal Nos. 2016A7002 and 2016B7002.

Conflicts of Interest: The authors declare no conflict of interest.

References

1. MacSleyne, J.; Uchic, M.D.; Simmons, J.P.; De Graef, M. Three-dimensional analysis of secondary γ/γ' precipitates in René-88 DT and UMF-20 superalloys. *Acta Mater.* **2009**, *57*, 6251–6267. [CrossRef]
2. Rowenhorst, D.J.; Lewis, A.C.; Spanos, G. Three-dimensional analysis of grain topology and interface curvature in a β-titanium alloy. *Acta Mater.* **2010**, *58*, 5511–5519. [CrossRef]
3. Adams, B.L. Orientation imaging microscopy: Emerging and future applications. *Ultramicroscopy* **1997**, *67*, 11–17. [CrossRef]
4. Vermeij, T.; De Graef, M.; Hoefnagels, J. Demonstrating the potential of accurate absolute cross-grain stress and orientation correlation using electron backscatter diffraction. *Scr. Mater.* **2019**, *162*, 266–271. [CrossRef]
5. Larson, B.C.; Yang, W.; Ice, G.E.; Budai, J.D.; Tischler, J.Z. Three-dimensional X-ray structural microscopy with submicrometre resolution. *Nature* **2002**, *415*, 887–890. [CrossRef] [PubMed]
6. Ice, G.E.; Larson, B.C.; Yang, W.; Budai, J.D.; Tischler, J.Z.; Pang, J.W.L.; Barabash, R.I.; Liu, W. Polychromatic X-ray microdiffraction studies of mesoscale structure and dynamics. *J. Synchrotron Rad.* **2005**, *12*, 155–162. [CrossRef]
7. Levine, L.E.; Larson, B.C.; Yang, W.; Kassner, M.E.; Tischler, J.Z.; Delos-Reyes, M.A.; Fields, R.J.; Liu, W. X-ray microbeam measurements of individual dislocation cell elastic strains in deformed single-crystal copper. *Nat. Mater.* **2006**, *5*, 619–622. [CrossRef]
8. Barabash, R.I.; Ice, G.E.; Liu, W.; Barabash, O.M. Polychromatic microdiffraction characterization of defect gradients in severely deformed materials. *Micron* **2009**, *40*, 28–36. [CrossRef]
9. Larson, B.C.; Levine, L.E. Submicrometre-resolution polychromatic three-dimensional X-ray microscopy. *J. Appl. Crystallogr.* **2013**, *46*, 153–164. [CrossRef]
10. Guo, Y.; Collins, D.M.; Tarleton, E.; Hofmann, F.; Tischler, J.; Liu, W.; Xu, R.; Wilkinson, A.J.; Britton, T.B. Measurements of stress fields near a grain boundary: Exploring blocked arrays of dislocations in 3D. *Acta Mater.* **2015**, *96*, 229–236. [CrossRef]

11. Zhang, C.; Li, H.; Eisenlohr, P.; Liu, W.; Boehlert, C.J.; Crimp, M.A.; Bieler, T.R. Effect of realistic 3D microstructure in crystal plasticity finite element analysis of polycrystalline Ti-5Al-2.5Sn. *Int. J. Plast.* **2015**, *69*, 21–35. [CrossRef]
12. Yu, T.; Du, Y.; Fan, G.; Barabash, R.; Juul Jensen, D.; Zhang, Y. In Situ synchrotron X-ray micro-diffraction investigation of elastic strains in laminated Ti-Al composites. *Metals* **2021**, *11*, 668. [CrossRef]
13. Zhang, Y.; Yu, T.; Xu, R.; Thorborg, J.; Liu, W.; Tischler, J.; Godfrey, A.; Juul Jensen, D. Local residual stresses and microstructure within recrystallizing grains in iron. *Mater. Charact.* **2022**, *191*, 112113. [CrossRef]
14. King, A.; Johnson, G.; Engelberg, D.; Ludwig, W.; Marrow, J. Observations of intergranular stress corrosion cracking in a grain-mapped polycrystal. *Science* **2008**, *321*, 382–385. [CrossRef]
15. Ludwig, W.; Schmidt, S.; Lauridsen, E.M.; Poulsen, H.F. X-ray diffraction contrast tomography: A novel technique for three-dimensional grain mapping of polycrystals. I. Direct beam case. *J. Appl. Crystallogr.* **2008**, *41*, 302–309. [CrossRef]
16. Johnson, G.; King, A.; Honnicke, M.G.; Marrow, J.; Ludwig, W. X-ray diffraction contrast tomography: A novel technique for three-dimensional grain mapping of polycrystals. II. The combined case. *J. Appl. Crystallogr.* **2008**, *41*, 310–318. [CrossRef]
17. Ludwig, W.; Reischig, P.; King, A.; Herbig, M.; Lauridsen, E.M.; Johnson, G.; Marrow, T.J.; Buffiere, J.Y. Three-dimensional grain mapping by x-ray diffraction contrast tomography and the use of Friedel pairs in diffraction data analysis. *Rev. Sci. Instrum.* **2009**, *80*, 033905. [CrossRef]
18. Reischig, P.; King, A.; Nervo, L.; Viganó, N.; Guilhem, Y.; Palenstijn, W.J.; Batenburg, K.J.; Preuss, M.; Ludwig, W. Advances in X-ray diffraction contrast tomography: Flexibility in the setup geometry and application to multiphase materials. *J. Appl. Crystallogr.* **2013**, *46*, 297–311. [CrossRef]
19. Nervo, L.; King, A.; Fitzner, A.; Ludwig, W.; Preuss, M. A study of deformation twinning in a titanium alloy by X-ray diffraction contrast tomography. *Acta Mater.* **2016**, *105*, 417–428. [CrossRef]
20. Reischig, P.; Ludwig, W. Three-dimensional reconstruction of intragranular strain and orientation in polycrystals by near-field X-ray diffraction. *Curr. Opin. Solid State Mater. Sci.* **2020**, *24*, 100851. [CrossRef]
21. Margulies, L.; Winther, G.; Poulsen, H.F. In-situ measurement of grain rotation during deformation of polycrystals. *Science* **2001**, *291*, 2392–2394. [CrossRef] [PubMed]
22. Poulsen, H.F.; Nielsen, S.F.; Lauridsen, E.M.; Schmidt, S.; Suter, R.M.; Lienert, U.; Margulies, L.; Lorentzen, T.; Juul Jensen, D. Three-dimensional maps of grain boundaries and the stress state of individual grains in polycrystals and powders. *J. Appl. Crystallogr.* **2001**, *34*, 751–756. [CrossRef]
23. Lauridsen, E.M.; Schmidt, S.; Suter, R.M.; Poulsen, H.F. Tracking: A method for structural characterization of grains in powders or polycrystals. *J. Appl. Crystallogr.* **2001**, *34*, 744–750. [CrossRef]
24. Nielsen, S.F.; Lauridsen, E.M.; Juul Jensen, D.; Poulsen, H.F. A three-dimensional X-ray diffraction microscope for deformation studies of polycrystals. *Mater. Sci. Eng. A* **2001**, *319–321*, 179–181. [CrossRef]
25. Offerman, S.E.; van Dijk, N.H.; Sietsma, J.; Grigull, S.; Lauridsen, E.M.; Margulies, L.; Poulsen, H.F.; Rekveldt, M.T.; van der Zwaag, S. Grain nucleation and growth during phase transformations. *Science* **2002**, *298*, 1003–1005. [CrossRef] [PubMed]
26. Margulies, L.; Lorentzen, T.; Poulsen, H.F.; Leffers, T. Strain tensor development in a single grain in the bulk of a polycrystal under loading. *Acta Mater.* **2002**, *50*, 1771–1779. [CrossRef]
27. Poulsen, H.F.; Fu, X. Generation of grain maps by an algebraic reconstruction technique. *J. Appl. Crystallogr.* **2003**, *36*, 1062–1068. [CrossRef]
28. Poulsen, H.F.; Margulies, L.; Schmidt, S.; Winther, G. Lattice rotations of individual bulk grains: Part I: 3D X-ray characterization. *Acta Mater.* **2003**, *51*, 3821–3830. [CrossRef]
29. Fu, X.; Poulsen, H.F.; Schmidt, S.; Nielsen, S.F.; Lauridsen, E.M.; Juul Jensen, D. Non-destructive mapping of grains in three dimensions. *Scr. Mater.* **2003**, *49*, 1093–1096. [CrossRef]
30. Schmidt, S.; Nielsen, S.F.; Gundlach, C.; Margulies, L.; Huang, X.; Juul Jensen, D. Watching the growth of bulk grains during recrystallization of deformed metals. *Science* **2004**, *305*, 229–232. [CrossRef]
31. Lauridsen, E.M.; Schmidt, S.; Nielsen, S.F.; Margulies, L.; Poulsen, H.F.; Juul Jensen, D. Non-destructive characterization of recrystallization kinetics using three-dimensional X-ray diffraction microscopy. *Scr. Mater.* **2006**, *55*, 51–56. [CrossRef]
32. Fu, X.; Knudsen, E.; Poulsen, H.F.; Herman, G.T.; Carvalho, B.M.; Liao, H.Y. Optimized algebraic reconstruction technique for generation of grain maps based on three-dimensional x-ray diffraction (3DXRD). *Opt. Eng.* **2006**, *45*, 1–9. [CrossRef]
33. Alpers, A.; Poulsen, H.F.; Knudsen, E.; Herman, G.T. A discrete tomography algorithm for improving the quality of three-dimensional X-ray diffraction grain maps. *J. Appl. Crystallogr.* **2006**, *39*, 582–588. [CrossRef]
34. Offerman, S.E.; van Dijk, N.H.; Sietsma, J.; Lauridsen, E.M.; Margulies, L.; Grigull, S.; Poulsen, H.F.; van der Zwaag, S. Phase transformations in steel studied by 3DXRD microscopy. *Nucl. Instrum. Methods Phys. Res. Sect. B* **2006**, *246*, 194–200. [CrossRef]
35. Rodek, L.; Poulsen, H.F.; Knudsen, E.; Herman, G.T. A stochastic algorithm for reconstruction of grain maps of moderately deformed specimens based on X-ray diffraction. *J. Appl. Crystallogr.* **2007**, *40*, 313–321. [CrossRef]
36. Jimenez-Melero, E.; van Dijk, N.H.; Zhao, L.; Sietsma, J.; Offerman, S.E.; Wright, J.P.; van der Zwaag, S. Characterization of individual retained austenite grains and their stability in low-alloyed TRIP steels. *Acta Mater.* **2007**, *55*, 6713–6723. [CrossRef]
37. Schmidt, S.; Olsen, U.L.; Poulsen, H.F.; Sørensen, H.O.; Lauridsen, E.M.; Margulies, L.; Maurice, C.; Juul Jensen, D. Direct observation of 3-D grain growth in Al–0.1% Mn. *Scr. Mater.* **2008**, *59*, 491–494. [CrossRef]
38. Oddershede, J.; Schmidt, S.; Poulsen, H.F.; Sørensen, H.O.; Wright, J.; Reimers, W. Determining grain resolved stresses in polycrystalline materials using three-dimensional X-ray diffraction. *J. Appl. Crystallogr.* **2010**, *43*, 539–549. [CrossRef]

39. Juul, N.Y.; Winther, G.; Dale, D.; Koker, M.K.A.; Shade, P.; Oddershede, J. Elastic interaction between twins during tensile deformation of austenitic stainless steel. *Scr. Mater.* **2016**, *120*, 1–4. [CrossRef]
40. Renversade, L.; Borbély, A. Evaluation of grain-average stress tensor in a tensile-deformed Al–Mn polycrystal by high-energy X-ray diffraction. *J. Appl. Crystallogr.* **2017**, *50*, 1144–1157. [CrossRef]
41. Abdolvand, H.; Wright, J.; Wilkinson, A.J. Strong grain neighbour effects in polycrystals. *Nat. Commun.* **2018**, *9*, 171. [CrossRef] [PubMed]
42. Oddershede, J.; Camin, B.; Schmidt, S.; Mikkelsen, L.P.; Sørensen, H.O.; Lienert, U.; Poulsen, H.F.; Reimers, W. Measuring the stress field around an evolving crack in tensile deformed Mg AZ31 using three-dimensional X-ray diffraction. *Acta Mater.* **2012**, *60*, 3570–3580. [CrossRef]
43. Sedmák, P.; Pilch, J.; Heller, L.; Kopeček, J.; Wright, J.; Sedlák, P.; Frost, M.; Šittner, P. Grain-resolved analysis of localized deformation in nickel-titanium wire under tensile load. *Science* **2016**, *353*, 559–562. [CrossRef]
44. Suter, R.M.; Hennessy, D.; Xiao, C.; Lienert, U. Forward modeling method for microstructure reconstruction using X-ray diffraction microscopy: Single-crystal verification. *Rev. Sci. Instrum.* **2006**, *77*, 123905. [CrossRef]
45. Hefferan, C.; Li, S.F.; Lind, J.; Lienert, U.; Rollett, A.D.; Wynblatt, P.; Suter, R. Statistics of high purity nickel microstructure from high energy X-ray diffraction microscopy. *Comput. Mater. Contin.* **2009**, *14*, 209–219.
46. Aydıner, C.C.; Bernier, J.V.; Clausen, B.; Lienert, U.; Tomé, C.N.; Brown, D.W. Evolution of stress in individual grains and twins in a magnesium alloy aggregate. *Phys. Rev. B* **2009**, *80*, 024113. [CrossRef]
47. Bernier, J.V.; Barton, N.R.; Lienert, U.; Miller, M.P. Far-field high-energy diffraction microscopy: A tool for intergranular orientation and strain analysis. *J. Strain Anal. Eng. Des.* **2011**, *46*, 527–547. [CrossRef]
48. Lienert, U.; Li, S.F.; Hefferan, C.M.; Lind, J.; Suter, R.M.; Bernier, J.V.; Barton, N.R.; Brandes, M.C.; Mills, M.J.; Miller, M.P.; Jakobsen, B.; et al. High-energy diffraction microscopy at the advanced photon source. *JOM* **2011**, *46*, 527–547. [CrossRef]
49. Li, S.F.; Lind, J.; Hefferan, C.M.; Pokharel, R.; Lienert, U.; Rollett, A.D.; Suter, R.M. Three-dimensional plastic response in polycrystalline copper *Via* Near-Field High-Energy X-ray Diffr. Microscopy. *J. Appl. Crystallogr.* **2012**, *45*, 1098–1108. [CrossRef]
50. Hefferan, C.M.; Lind, J.; Li, S.F.; Lienert, U.; Rollett, A.D.; Suter, R.M. Observation of recovery and recrystallization in high-purity aluminum measured with forward modeling analysis of high-energy diffraction microscopy. *Acta Mater.* **2012**, *60*, 4311–4318. [CrossRef]
51. Li, S.F.; Suter, R.M. Adaptive reconstruction method for three-dimensional orientation imaging. *J. Appl. Crystallogr.* **2013**, *46*, 512–524. [CrossRef]
52. Pokharel, R.; Lind, J.; Kanjarla, A.K.; Lebensohn, R.A.; Li, S.F.; Kenesei, P.; Suter, R.M.; Rollett, A.D. Polycrystal plasticity: Comparison between grain-scale observations of deformation and simulations. *Annu. Rev. Condens. Matter Phys.* **2014**, *5*, 317–346. [CrossRef]
53. Wang, L.; Lind, J.; Phukan, H.; Kenesei, P.; Park, J.-S.; Suter, R.M.; Beaudoin, A.J.; Bieler, T.R. Mechanical twinning and detwinning in pure Ti during loading and unloading—An in situ high-energy X-ray diffraction microscopy study. *Scr. Mater.* **2014**, *92*, 35–38. [CrossRef]
54. Lind, J.; Li, S.F.; Pokharel, R.; Lienert, U.; Rollett, A.D.; Suter, R.M. Tensile twin nucleation events coupled to neighboring slip observed in three dimensions. *Acta Mater.* **2014**, *76*, 213–220. [CrossRef]
55. Spear, A.D.; Li, S.F.; Lind, J.f.; Suter, R.M.; Ingraffea, A.R. Three-dimensional characterization of microstructurally small fatigue-crack evolution using quantitative fractography combined with post-mortem X-ray tomography and high-energy X-ray diffraction microscopy. *Acta Mater.* **2014**, *76*, 413–424. [CrossRef]
56. Miller, M.P.; Dawson, P.R. Understanding local deformation in metallic polycrystals using high energy X-rays and finite elements. *Curr. Opin. Solid State Mater. Sci.* **2014**, *18*, 286–299. [CrossRef]
57. Shade, P.A.; Blank, B.; Schuren, J.C.; Turner, T.J.; Kenesei, P.; Goetze, K.; Suter, R.M.; Bernier, J.V.; Li, S.F.; Lind, J; et al. A rotational and axial motion system load frame insert for in situ high energy X-ray studies. *Rev. Sci. Instrum.* **2015**, *86*, 093902. [CrossRef]
58. Schuren, J.C.; Shade, P.A.; Bernier, J.V.; Li, S.F.; Blank, B.; Lind, J.; Kenesei, P.; Lienert, U.; Suter, R.M.; Turner, T.J.; et al. New opportunities for quantitative tracking of polycrystal responses in three dimensions. *Curr. Opin. Solid State Mater. Sci.* **2015**, *19*, 235–244. [CrossRef]
59. Wong, S.L.; Obstalecki, M.; Miller, M.P.; Dawson, P.R. Stress and deformation heterogeneity in individual grains within polycrystals subjected to fully reversed cyclic loading. *J. Mech. Phys. Solids.* **2015**, *79*, 157–185. [CrossRef]
60. Turner, T.; Shade, P.; Bernier, J.; Li, S.F.; Schuren, J.; Kenesei, P.; Suter, R.; Almer, J. Crystal plasticity model validation using combined high-energy diffraction microscopy data for a Ti-7Al specimen. *Metall. Mater. Trans. A Phys. Metall. Mater. Sci.* **2016**, *48*, 627–647. [CrossRef]
61. Pagan, D.C.; Shade, P.A.; Barton, N.R.; Park, J.-S.; Kenesei, P.; Menasche, D.B.; Bernier, J.V. Modeling slip system strength evolution in Ti-7Al informed by in-situ grain stress measurements. *Acta Mater.* **2017**, *128*, 406–417. [CrossRef]
62. Guillen, D.P.; Pagan, D.C.; Getto, E.M.; Wharry, J.P. In situ tensile study of PM-HIP and wrought 316L stainless steel and Inconel 625 alloys with high energy diffraction microscopy. *Mater. Sci. Eng. A* **2018**, *738*, 380–388. [CrossRef]
63. Naragani, D.; Sangid, M.D.; Shade, P.A.; Schuren, J.C.; Sharma, H.; Park, J.-S.; Kenesei, P.; Bernier, J.V.; Turner, T.J.; Parr, I. Investigation of fatigue crack initiation from a non-metallic inclusion via high energy X-ray diffraction microscopy. *Acta Mater.* **2017**, *137*, 71–84. [CrossRef]

64. Pagan, D.C.; Beaudoin, A.J. Utilizing a novel lattice orientation based stress characterization method to study stress fields of shear bands. *J. Mech. Phys. Solids.* **2019**, *128*, 105–116. [CrossRef]
65. Shen, Y.-F.; Liu, H.; Suter, R.M. Voxel-based strain tensors from near-field high energy diffraction microscopy. *Curr. Opin. Solid State Mater. Sci.* **2020**, *24*, 100852. [CrossRef]
66. Naragani, D.; Shade, P.; Musinski, W.; Boyce, D.; Obstalecki, M.; Pagan, D.; Bernier, J.; Beaudoin, A. Interpretation of intragranular strain fields in high-energy synchrotron X-ray experiments via finite element simulations and analysis of incompatible deformation. *Mater. Des.* **2021**, *210*, 110053. [CrossRef]
67. Hayashi, Y.; Hirose, Y.; Setoyama, D. In situ three-dimensional orientation mapping in plastically-deformed polycrystalline iron by three-dimensional X-ray diffraction. *Mater. Sci. Forum* **2014**, *777*, 118–123. [CrossRef]
68. Bonnin, A.; Wright, J.P.; Tucoulou, R.; Palancher, H. Impurity precipitation in atomized particles evidenced by nano X-ray diffraction computed tomography. *Appl. Phys. Lett.* **2014**, *105*, 084103. [CrossRef]
69. Hayashi, Y.; Hirose, Y.; Seno, Y. Polycrystal orientation mapping using scanning three-dimensional X-ray diffraction microscopy. *J. Appl. Crystallogr.* **2015**, *48*, 1094–1101. [CrossRef]
70. Hayashi, Y.; Hirose, Y.; Seno, Y. Scanning three-dimensional X-ray diffraction microscopy using a high-energy microbeam. *AIP Conf. Proc.* **2016**, *1741*, 050024.
71. Hayashi, Y.; Setoyama, D.; Seno, Y. Scanning three-dimensional X-ray diffraction microscopy with a high-energy microbeam at SPring-8. *Mater. Sci. Forum* **2017**, *905*, 157–164. [CrossRef]
72. Hektor, J.; Hall, S.; Henningsson, N.; Engqvist, J.; Ristinmaa, M.; Lenrick, F.; Wright, J. Scanning 3DXRD measurement of grain growth, stress, and formation of Cu_6Sn_5 around a tin whisker during heat treatment. *Materials* **2019**, *12*, 446. [CrossRef] [PubMed]
73. Hayashi, Y.; Setoyama, D.; Hirose, Y.; Yoshida, T.; Kimura, H. Intragranular three-dimensional stress tensor fields in plastically deformed polycrystals. *Science* **2019**, *366*, 1492–1496. [CrossRef] [PubMed]
74. Henningsson, N.A.; Hall, S.A.; Wright, J.P.; Hektor, J. Reconstructing intragranular strain fields in polycrystalline materials from scanning 3DXRD data. *J. Appl. Crystallogr.* **2020**, *53*, 314–325. [CrossRef] [PubMed]
75. Wright J.P.; Giacobbe, C.; Majkut, M. New opportunities at the materials science beamline at ESRF to exploit high energy nano-focus X-ray beams. *Curr. Opin. Solid State Mater. Sci.* **2020**, *24*, 100818. [CrossRef]
76. Henningsson, A.; Hendriks, J. Intragranular strain estimation in far-field scanning X-ray diffraction using a Gaussian process. *J. Appl. Crystallogr.* **2021**, *54*, 1057–1070. [CrossRef]
77. Li, W.; Sharma, H.; Peter, K.; Ravi, S.; Sehitoglu, H.; Bucsek, A. Resolving intragranular stress fields in plastically deformed titanium using point-focused high-energy diffraction microscopy. *J. Mater. Res.* **2023**, *38*, 165–178. [CrossRef]
78. Nielsen, S.F.; Wolf, A.; Poulsen, H.F.; Ohler, M.; Lienert, U.; Owen, R.A. A conical slit for three-dimensional XRD mapping. *J. Synchrotron Rad.* **2000**, *7*, 103–109. [CrossRef]
79. Wright, J. ImageD11. 2021. Available online: https://github.com/FABLE-3DXRD/ImageD11/ (accessed on 7 July 2023).
80. Wright, J. PolyXSim. 2019. Available online: https://github.com/FABLE-3DXRD/PolyXSim (accessed on 7 July 2023).

quantum beam science

MDPI

Article

Scanning Three-Dimensional X-ray Diffraction Microscopy with a Spiral Slit

Yujiro Hayashi †, Daigo Setoyama, Kunio Fukuda, Katsuharu Okuda, Naoki Katayama and Hidehiko Kimura *

Toyota Central R&D Labs., Inc., 41-1, Yokomichi, Nagakute 480-1192, Aichi, Japan
* Correspondence: hdkimura@mosk.tytlabs.co.jp
† Current address: RIKEN SPring-8 Center, 1-1-1 Kouto, Sayo-cho, Sayo-gun 679-5148, Hyogo, Japan.

Abstract: Recently, nondestructive evaluation of the stresses localized in grains was achieved for plastically deformed low-carbon steel using scanning three-dimensional X-ray diffraction (S3DXRD) microscopy with a conical slit. However, applicable metals and alloys were restricted to a single phase and evaluated stress was underestimated due to the fixed Bragg angles of the conical slit optimized to αFe. We herein propose S3DXRD with a rotating spiral slit adaptable to various metals and alloys and accurate stress evaluation with sweeping Bragg angles. Validation experiments with a 50-keV X-ray microbeam were conducted for low-carbon steel as a body-centered cubic (BCC) phase and pure Cu as a face-centered cubic (FCC) phase. As a result of orientation mapping, polygonal grain shapes and clear grain boundaries were observed for both BCC and FCC metals. Thus, it was demonstrated that S3DXRD with a rotating spiral slit will be applicable to various metals and alloys, multiphase alloys, and accurate stress evaluation using a X-ray microbeam with a higher photon energy within an energy range determined by X-ray focusing optics. In principle, this implies that S3DXRD becomes applicable to larger and thicker metal and alloy samples instead of current miniature test or wire-shaped samples if a higher-energy X-ray microbeam is available.

Keywords: 3DXRD; spiral slit; orientation

1. Introduction

The localized stresses in the polycrystalline metals and alloys produced due to fabrication processes and fatigue in service are crucial study objects in engineering and in many industries. Three-dimensional X-ray diffraction (3DXRD) microscopy (also called high-energy diffraction microscopy or HEDM) with synchrotron-based high-energy X-rays allows us to nondestructively evaluate grain-resolved stresses called type II stresses [1–13]. The mapping of type II stresses has successfully led to the visualization of macroscopic (type I) stress fields, such as crack-tip stress fields, and to the deviation of type II stresses from type I stresses [14,15]. Three-dimensional type II stress mapping combined with computed tomography has demonstrated that fatigue crack initiations from inclusions occur at grain boundaries, showing high stress gradients [16].

The determination of type II stresses is based on the detection of diffracted beams from multiple grains as diffraction spots on area detectors. The diffraction spots can be assigned to multiple grains with known crystallographic parameters, which is called multigrain indexing. If multiple diffraction spots per grain are assigned, orientation is determined for individual grains. Type II stresses are determined as follows. First, lattice parameters are calculated for individual grains. The lattice parameters can be converted into an elastic strain tensor in a crystallographic coordinate system if stress-free lattice parameters are defined using reference samples, e.g., annealed powders. An elastic strain tensor is converted into a stress tensor in a crystallographic coordinate system through elastic stiffness constants. Then, the stress tensor is converted into a stress tensor in the sample coordinate system through orientation. Three-dimensional mapping of determined stress tensors is achieved by reconstructing the positions of individual grains, including or excluding grain shapes with spatial resolutions, which is called grain mapping. In 3DXRD with monochromatic X-rays, multiple diffraction spots per grain are acquired by rotating samples, and grain maps are reconstructed from diffraction spots at near fields detected with a high-spatial-resolution detector.

Similar to type II stresses compared with type I stresses, intragranular (type III) stresses may deviate from grain-averaged type II stresses [17–24]. Recently, the evaluation of type III stresses has been achieved using a 3DXRD modality termed scanning 3DXRD (also called point-focused HEDM) with a monochromatic point-focused (pencil) beam [25–27]. In scanning 3DXRD, a voxel-resolved stress tensor with a voxel size smaller than a grain size can be obtained by illuminating a pencil beam and rotating a sample. Grain mapping with grain boundaries can be achieved by a three-dimensional scan of a sample, including sample rotations at the expense of measurement time [28]. Unsurprisingly, voxel-resolved stresses showed deviations from grain-averaged stresses in a plastically deformed metal and alloy [27,29]. It was also shown that deviation exceeded macroscopic tensile strength and voxel-resolved stresses were in highly triaxial stress states even under uniform elongation. Reconstructions with high fidelity have revealed the expectations of the larger deviations of voxel-resolved stresses from grain-averaged stresses [30,31].

Very recently, a high-spatial-resolution modality of 3DXRD, called high-resolution 3DXRD (HR-3DXRD), has been proposed [32]. In HR-3DXRD, if diffracted beams from submicrometer-sized crystallites or subgrains in a deformed grain are detected as diffraction spots by locating a high-resolution area detector at a sample-to-detector distance with a balance of real and reciprocal resolutions, multiple diffraction spots are assigned for individual crystallites or subgrains in a grain. Therefore, type III stress can be evaluated for plastically deformed metals and alloys without the time-consuming three-dimensional scans.

The limitations of the type II and type III stress evaluation methods, including HR-3DXRD, come from the overlap of diffraction spots, which gives rise to stress evaluation errors. Diffraction spot overlaps are caused by the existence of a large number of grains in an illuminated volume and the spread of diffraction spots brought due to mosaicity, subgrains, and intragranular misorientations. Therefore, 3DXRD-based experiments are limited to small samples, such as miniature test or wire-shaped samples. Sample miniaturization causes size effects that lead to different mechanical material behaviors, which depend on surface conditions, sample fabrication methods, microstructure inhomogeneity, grain size, etc. [33]. In fact, industrial standardizations of miniature test samples are unavailable. The limitation on applicable sample sizes comes from the overlap of diffraction spots due to a large number of grains and subgrains rather than the availability of synchrotron-based X-rays with higher photon energy.

The introduction of conical slits [34] to 3DXRD-based experiments is meaningful from the viewpoint of applicable sample sizes. Conical slits can significantly reduce the overlap of diffraction spots by shielding detection instruments from the diffracted beams of many grains, except for the diffracted beams from the grains inside regions of interest (i.e., gauge volumes). 3DXRD with a conical slit was demonstrated to observe crystallographic rotations of individual grains in mm-sized pure aluminum during plastic deformation [35,36]. Scanning 3DXRD with a conical slit was demonstrated to evaluate type III stress in plastically deformed mm-sized low-carbon steel [29]. 3DXRD-based techniques, including HR-3DXRD, can be combined with conical slits and then applied to larger and thicker metallic samples if higher-energy X-rays are available. However, the conical slit in such 3DXRD-based experiments needs to be optimized for a single crystalline phase, because thicker metallic samples require synchrotron-based X-rays with the photon energy of almost the upper limit of a tunable X-ray energy range restricted by X-ray sources, beamline optics, and focusing optics. Therefore, it is difficult to apply 3DXRD-based techniques with conical slits to larger and thicker samples of various metals and alloys.

The limitation of the conical slit is caused by fixed Bragg angles 2θ. A solution to the limitation is the use of a 2θ-sweepable rotating spiral slit [37]. Using the rotating spiral slit, diffraction spots could be detected independently of crystalline phases from grains inside a gauge volume in a thick metallic sample. Here, we introduced a rotating spiral slit to scanning 3DXRD. As the first step of scanning 3DXRD with a spiral slit, we demonstrated orientation mapping for non-deformed single-phase metals without significant intragranular misorientations. The orientation mapping for metals of body-centered cubic (BCC) and face-centered cubic (FCC) lattices was achieved by scanning 3DXRD with a compact rotating spiral slit and a 50-keV X-ray microbeam. The X-ray energy was determined as the upper limit of the tunable energy range determined by focusing optics.

2. Materials and Methods

2.1. Data Acquisition

In scanning 3DXRD, a monochromatic X-ray pencil beam with a beam size smaller than a grain size illuminates a polycrystalline sample. While the sample rotates around the z-axis, diffracted beams from multiple grains are detected as diffraction spots on a far-field area detector, as shown in Figure 1a, where xyz is the laboratory coordinate system and the incoming pencil beam is parallel to the y-axis. Diffraction images from the far-field detector are continuously acquired every ω step from $\omega = 0°$ to $\omega = \omega_{max}$, where ω is the rotation angle of the sample around the z-axis. Because of tomographic reconstruction stated below with a computed tomography-like geometry, 180° at least is needed for ω_{max}. After this ω scan, the sample is translated by ΔX in the x direction using a translation stage located under an ω rotation stage. Although the X scan step ΔX should be matched to the beam size, ΔX slightly larger than the beam size is acceptable for scan time saving because slight sampling errors are negligible as long as ΔX is much smaller than a grain size. The ω scan is repeated from $X = -X_m$ to $X = X_m$, where X is the sample translation in the x direction, and the incoming microbeam penetrated the ω rotation axis at $X = 0$. From this ω-X scan data, an orientation map with a spatial resolution of $\sim 2\Delta X$ and a field-of-view (FoV) with a diameter of $2X_m$ in the xy plane can be reconstructed using the scanning 3DXRD method [28].

2.2. Reconstruction

Here, we summarize the reconstruction procedures of the scanning 3DXRD method [28] as follows.

- *Extraction.* Certain specific diffraction images are extracted from the ω-X scan data. The extracted specific diffraction images correspond to the acquired images when the incoming beam penetrates an arbitrary point Q. In this condition, only the point Q is always illuminated by the incoming beam during the ω_{max} rotations. A region with a center of Q and a diameter of $\sim 2\Delta X$ is also illuminated during the ω_{max} rotations due to the beam size, which corresponds to a spatial resolution to determine orientation.
- *Multigrain indexing.* Multiple grains can be produced as candidate grains that occupies the point Q by applying multigrain indexing for the extracted diffraction images. Orientations and N are determined for each candidate grain, where N is the number of detected diffraction spots for a grain. If the sample consists of multiple crystalline phases, multigrain indexing is conducted for each phase.
- *Normalization.* A normalization factor, M, for the dependency of N on orientation and crystalline phases is calculated, where M is the theoretically expected maximum of N. The normalization factor M is calculated considering not only a sample-to-detector distance and the active area size of the detector but also non-detectable (shadow) areas due to the holder of a beam stop, the rotation mechanism of a spiral slit, etc.
- *Repetition.* An arbitrary point can be selected anywhere within FoV with a diameter of $2X_m$ and a center matched with an ω rotation center in the xy plane. The above extraction, multigrain indexing, and normalization are repeated for all of voxels within FoV. The voxel size is set to ΔX for simplicity.

Thus, we obtained an orientation map with multiple candidates of orientation per voxel with a spatial resolution of $\sim 2\Delta X$ and FoV of a diameter of $2X_{\mathrm{m}}$.

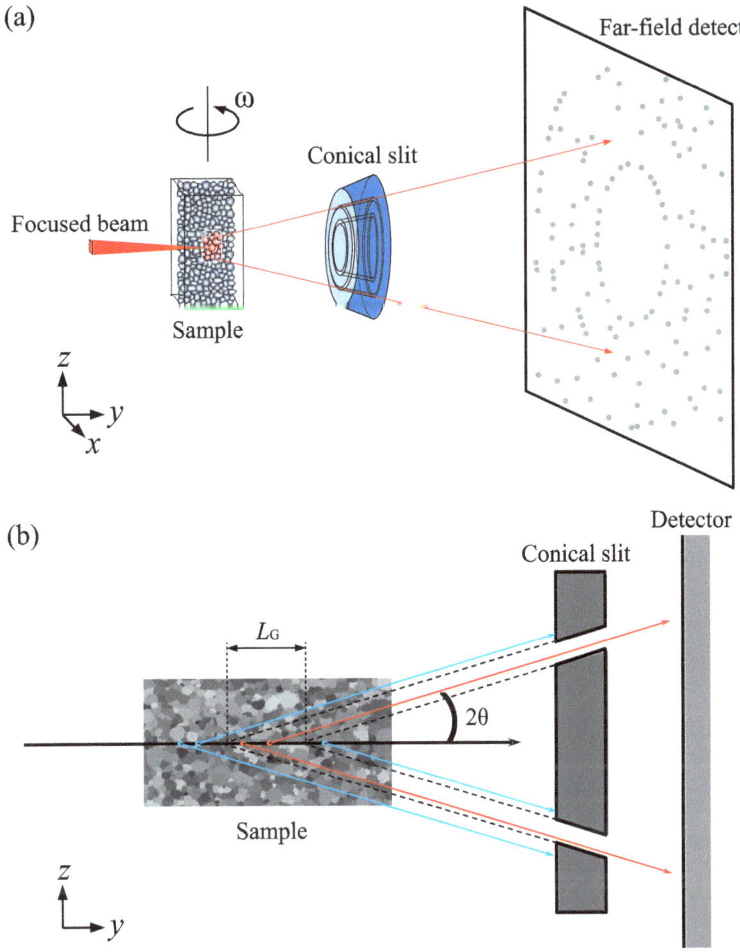

Figure 1. (**a**) Schematic of the experimental setup of scanning 3DXRD with a conical slit and (**b**) an illustration showing the role of the conical slit. Diffraction spots from multiple grains are detected through the conical slit with a far-field detector. The conical slit forms a gauge volume (L_G) in the sample, which allows us to reduce the overlap of diffraction spots by restricting the number of detectable grains. Generally, acceptable Bragg angles (2θ) need to be optimized for a single crystalline phase and use X-ray photon energy.

2.3. Postprocess

For ideal cases without significant experimental errors, an orientation map with single orientation per voxel is obtained by simply selecting orientation with the highest $N' = N/M$ among candidates at each voxel. For cases with experimental errors due to insufficient diffraction intensities, etc., a better orientation map can be obtained by applying postprocesses using the information on the orientation of neighboring voxels in addition to N'. Here, we proposed a simple postprocess as follows.

First, a voxel was randomly selected from an orientation map. Misorientations were evaluated between orientation with the highest N' in the selected voxel and orientation with first, second, and third highest N' in first neighboring voxels. If the misorientations were smaller than δ, the first neighboring voxels and the selected voxel were regarded as voxels in the same grain. The same judgment was conducted with misorientations between the first and second neighboring voxels. These steps were repeated until no new neighboring voxels were found. When the number of voxels regarded as voxels in the same grain was larger than a threshold, the voxels were extracted as a single grain and deleted from the original orientation map. Next, a voxel was randomly selected from the orientation map in which a single grain was deleted in the previous process. The same series of judgments with misorientations between neighboring voxels was conducted. Then, a next single grain might be extracted. The extractions of a single grain were repeated until no new grain was found. Thus, an orientation map with single orientation per voxel was obtained.

2.4. Role of a Spiral Slit

Multigrain indexing is based on the detection of diffraction spots from a polycrystalline sample. One of the limitations comes from the overlap of diffraction spots due to the detection of diffracted beams from a large number of strained grains. This situation mostly happens for thick strained metallic samples. We conquered the limitation of multigrain indexing-based techniques by developing scanning 3DXRD with a high-energy microbeam and a conical slit for single-phase alloys. Because a conical slit can form a gauge volume in a thick sample, a high-energy microbeam and the gauge volume allowed us to reduce the overlap of diffraction spots by restricting the number of detectable grains, as shown in Figure 1b. In addition, the conical slit can accept peak shifts $\Delta\theta$ of Bragg angles 2θ due to elastic strains for grains existing around the center of the gauge volume, as long as the conical slit can be located close to the sample. We achieved the evaluation of type III stress in 1-mm-sized plastically-deformed low-carbon ferrite steel using a microbeam with a photon energy of 50 keV and a conical slit designed for ferrite [29]. However, Bragg angles of the conical slit needed to be optimized for BCC ferrite and a photon energy of 50 keV. Although a conical slit designed for BCC can be applied to FCC if X-ray photon energy can be tuned, the scanning 3DXRD experiments with the conical slit were restricted to only BCC ferrite, because the X-ray photon energy range was limited by focusing optics and X-rays of the upper limit of 50 keV in the energy range were needed for the 1-mm-thick steel sample. Additionally, observed $\Delta\theta$ values were underestimated because it was difficult to locate the conical slit very close to the sample due to an in-situ stress rig, and then acceptable $\Delta\theta$ values were restricted.

A rotating spiral slit allowed us to overcome the obstacles of conical slits (Figure 2a). The rotating spiral slit has the functions of the sweeping of 2θ by rotating the spiral slit and the formation of a gauge volume (Figure 2b). Diffracted beams from various metals and alloys become acceptable by sweeping 2θ. Additionally, unlimited $\Delta\theta$ and strain are acceptable for all the grains in the gauge volume.

2.5. Experiments and Materials

Scanning 3DXRD experiments with a rotating spiral slit were performed using an undulator beamline BL33XU at SPring-8. Undulator radiation was monochromatized to 50 keV with a liquid nitrogen-cooled Si 311 double-crystal monochromator. The monochromator was calibrated with the K-edge of Cu. The monochromatic beam was focused into 1×1 µm using 400-mm-long Pt-coated Kirkpatrick-Baez mirrors with incident angles of <1.3 mrad. The focused microbeam was irradiated to a sample mounted on an ω rotation stage. The sample was rotated around the z-axis with a constant rotation speed of 1.2 °/s. Diffracted beams were detected through a rotating spiral slit using a flat-panel X-ray detector (2923NDT, Dexela, London, UK) with an active image area of 291 mm wide and 230 mm high and a sample-to-detector distance of 346 mm. The rotating spiral slit with a minimum aperture of 20 µm covered 2θ from 3.5° to 20°. Diffraction images were continuously recorded every 0.6° from 0° to $\omega_{max} = 180°$ during ω rotations. The exposure time of the images corresponds to 500 ms. Assuming that the mean grain size of the sample was more than 10 μm, the X scan step was set to $\Delta X = 1.2$ μm.

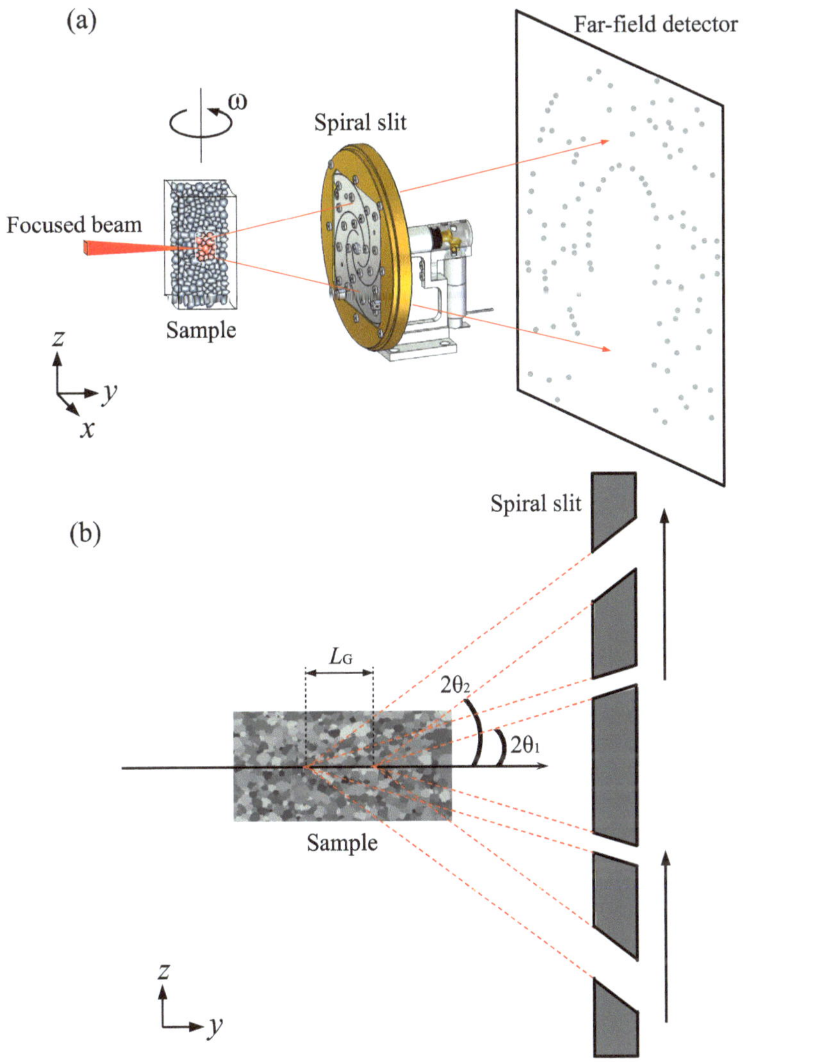

Figure 2. (**a**) Schematic of the experimental setup of scanning 3DXRD with a spiral slit and (**b**) an illustration showing the role of the spiral slit. Diffraction spots from multiple grains are detected through the spiral slit with a far-field detector. The spiral slit is rotated with a compact motor. The rotating spiral slit has the function of the sweeping of 2θ. Diffracted beams from multiple crystalline phases become acceptable by sweeping 2θ. In addition, $\Delta\theta$ and strain are acceptable, where $\Delta\theta$ is the deviation of Bragg angles due to strain. The spiral slit also forms a gauge volume (L_G) in the sample to reduce the overlap of diffraction spots by restricting the number of detectable grains. Scanning 3DXRD with the rotating spiral slit is applicable to multiple crystalline phases.

Single-phase BCC and FCC polycrystalline samples were prepared. The BCC sample was extracted from a commercially available cold-rolled steel sheet defined as SPCC according to the Japanese industrial standard [38]. The steel sheet with a thickness of 1 mm was cut using a wire cutter to have the shape of a tensile test piece with a width of 1 mm at a gauge section. The αFe sample with a tensile test piece shape was mounted on the ω stage with the longitudinal direction parallel to the z axis. Therefore, the cross-sectional area of the αFe sample in the observed xy plane was 1×1 mm^2. The FCC sample was cut from a commercially available pure Cu wire with a diameter of 1 mm. The Cu wire sample was mounted on the ω stage with the axial direction parallel to the z axis. Therefore, the cross-sectional area of the Cu sample in the observed xy plane was 0.79 mm^2.

2.6. Rotating Spiral Slit

A compact rotating spiral slit was fabricated so that the rotating spiral slit was located between the sample and the detector. The compact spiral slit with a diameter of 150 mm and a depth of 95 mm (external dimensions) was composed of parts of slit blades on a blade holder and a rotation mechanism with a compact motor, as shown in Figure 3. The blades were fabricated from three thin tungsten plates with a thickness of 1 mm and a width of 100 mm. Inclined through-grooves with a width of 0.5 mm were machined into the tungsten plates using a wire cutter, and inclined gaps were formed by stacking the three blades. Each blade had four spiral-shaped, 0.5-mm-wide grooves and a center bore, as shown in Figure 4a,b. The blades were mounted on a rigid holder made of brass with a center bore and a function of the alignment of each blade (Figure 4c). A pipe shaft was inserted into the center bore of the holder from the rear side and fixed to the holder. Using an appropriate hole shaft fit tolerance, the blades were mounted on the holder from the front side through the pipe shaft. Each blade was independently and slightly rotated around the y axis, where the pipe shaft was parallel to the y axis. Then, four spiral-shaped inclined gaps were formed. The gap widths, i.e., slit apertures, were designed as $L_G \tan 2\theta$ so that L_G was independent of 2θ, where L_G was the length of the gauge volume in the y direction. A minimum gap of 20 µm was formed at $2\theta = 3.5°$, corresponding to $L_G = \sim 330$ µm. After the alignment, the three blades were fixed to the pipe shaft through the holder. The pipe shaft was linked to a compact motor using reduction and bevel gears. The mechanical errors of the assembled rotating spiral slit were estimated to be about 15 µm and 35 µm at most for runout and deflection, respectively. The rotating spiral slit was mounted on a five-axis stage for three translations and two tilts and was located at a designed position with a working distance of 170 mm between the omega axis and surface of the first blade. The compact spiral slit was aligned so that the incoming beam passed through the internal bore of the pipe shaft (Figure 5a). The pipe shaft was rotated at 8400 rpm with an acceleration time of a few minutes (Figure 5b). The beam stop and detector were located at the downstream of the spiral slit. A reference powder sample in a capillary with a diameter similar to L_G was mounted at the rotation center of the omega stage with the longitudinal direction of the capillary parallel to the omega rotation axis. At first, Debye–Sherrer rings were partially observed from the rotating reference sample. Finally, the rotating spiral slit was aligned with the five-axis stage so that the whole Debye–Sherrer rings were observed.

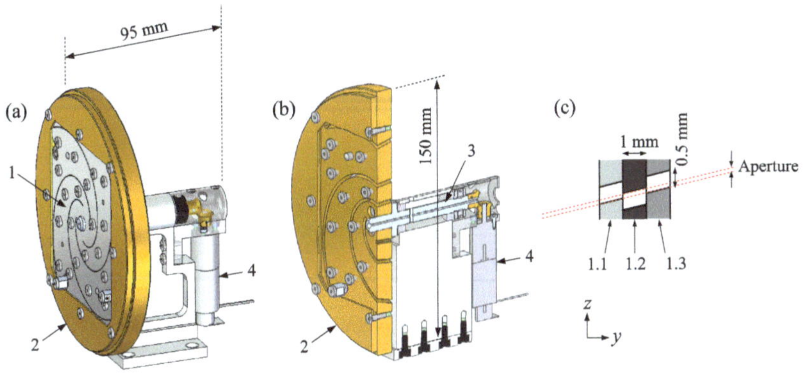

Figure 3. (**a**) External and (**b**) cross-sectional views of the design of the compact rotating spiral slit composed of tungsten blade plates (1), a blade holder (2), a pipe shaft (3), and a compact motor (4). (**c**) Schematic of the cross section of the tungsten blade plates with 0.5-mm-wide inclined through-grooves. An inclined slit aperture with a minimum gap of 20 μm is formed by aligning the first (1.1), second (1.2), and third (1.3) blade plates with a thickness of 1 mm.

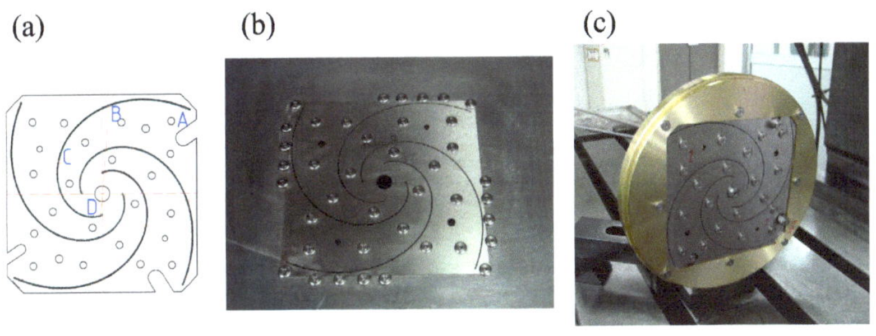

Figure 4. (**a**) Design drawing and (**b**) a photograph of a 100-mm-wide tungsten blade plate with four 0.5-mm-wide spiral-shaped inclined through-grooves. The inclination angles were ~20°, 14.5°, 9°, and 3.5° at parts A, B, C, and D, respectively. (**c**) Photograph of the assembled blades on the rigid holder made of brass.

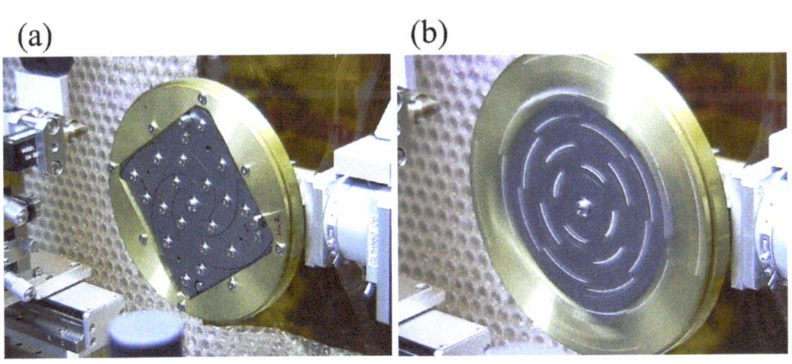

Figure 5. Photographs of (**a**) the assembled spiral slit located between the sample and detector and (**b**) the rotating spiral slit with a rotation speed of 8400 rpm.

3. Results and Discussion

The ω-X scans were conducted for X_m = 108 and 90 μm with measurement times of 8.5 and 7 h for the αFe and Cu samples, respectively. Orientations were analyzed with 110, 200, 211, 220, 310, and 222 reflections for αFe and 200, 220, 311, 400, 331, and 422 reflections for Cu. The postprocess for reconstruction was conducted with the misorientation parameter of $\delta = 2°$.

Figure 6 shows the orientation and N' maps obtained from the cold-rolled αFe steel sheet sample with a field-of-view of 216 μm in diameter. Each pixel with a pixel size of 1.2×1.2 μm has an orientation represented as the longitudinal direction of the tensile test shape sample in the basic triangle of the pole figure of BCC. In the orientation map, grains and grain boundaries are indicated by regions enclosed by the same orientations (or the same colors) and boundaries between different orientations (or different colors), respectively. Intragranular misorientations were estimated to be less than 1° for the grains. Uncolored white pixels imply reconstruction errors, not fine spheroidal cementite particles or voids. Polygonal grain shapes and clear grain boundaries are observed, which are commonly seen for annealed polycrystalline metals and alloys.

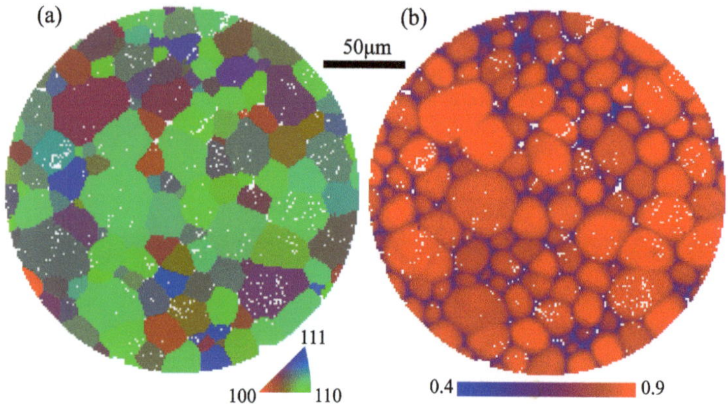

Figure 6. (**a**) Orientation and (**b**) N' maps obtained from the low-carbon steel sample. The field-of-view and the pixel size are 216 μm in diameter and 1.2×1.2 μm, respectively. Orientation is represented as the longitudinal direction of the tensile test shape sample in the basic triangle of the pole figure of BCC. The longitudinal direction is parallel to the z axis. Grains and grain boundaries are indicated by regions enclosed by the same orientations (or the same colors) and boundaries between different orientations (or different colors), respectively. Uncolored white pixels imply reconstruction errors.

The mean of N' and the range of M were estimated to be 0.70 and 81–86, respectively, from the N' map (Figure 6b). It can be seen that the N' values reach about 0.9 at the centers of grains, which indicates that about 90% of the theoretically expected number of diffraction spots per grain were detected. This result implies that the compact rotating spiral slit functioned and its mechanical errors were acceptable for orientation mapping.

Figure 7 shows orientation and N' maps obtained from the Cu wire sample with a field-of-view of 180 μm in diameter. In the orientation maps, the orientation is represented as the axial and radial directions in the basic triangle of the pole figure of FCC because of the wire texture. Although N' of small grains did not reach 0.9, it is obvious that polygonal grain shapes and grain boundaries were also successfully reconstructed for the FCC sample. Thus, the reconstruction of BCC/BCC and FCC/FCC grain boundaries was achieved for the BCC and FCC respective samples using a reconstruction parameter, N', independent of crystalline phases and the same experimental setup. This implies that BCC/FCC grain boundaries can be reconstructed with similar completeness.

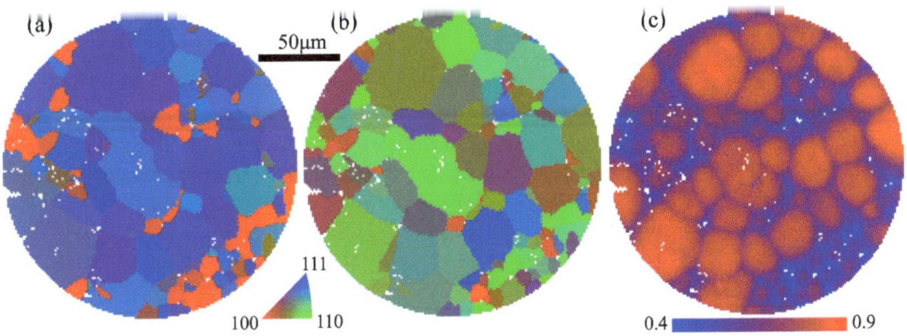

Figure 7. (**a**,**b**) Orientation and (**c**) N' maps obtained from the Cu wire sample. The field-of-view and the pixel size are 180 μm in diameter and 1.2 × 1.2 μm, respectively. Orientation is represented as the (**a**) axial and (**b**) radial directions of the wire sample in the basic triangle of the pole figure of FCC. The axial direction is parallel to the z axis. Uncolored white pixels imply reconstruction errors.

Orientation mapping was achieved for different metals by sweeping 2θ with a rotating spiral slit. In addition to orientation mapping, the spiral slit could let us evaluate type II and III stresses more quantitatively compared with conical slits, as $\Delta\theta$ is also acceptable by sweeping 2θ. The evaluation of type II and III stresses is interesting, especially for the multiphase alloys consisting of grains with different Young's moduli and yield strengths, such as TRIP steel, which contains BCC and FCC grains. The reconstruction of BCC/BCC, FCC/FCC, and BCC/FCC grain boundaries is important, as stress redistributions for BCC and FCC grains depend on neighboring grains. Scanning 3DXRD with a spiral slit is especially suitable for cases in which both grain mapping and type II and III stress evaluations are necessary.

Historically, a conical slit was reported in 2000 [34] and applied to 3DXRD in 2003 [35]. A rotating spiral slit and scanning 3DXRD were reported in 2014 [37] and 2015 [28]. Eventually, a rotating spiral slit has been applied to scanning 3DXRD in this work (2023). Although we used 1-mm-thick non-deformed metallic samples without significant intragranular misorientation, scanning 3DXRD with a rotating spiral slit has been successfully demonstrated as the first step. In principle, the method will be applicable to plastically-deformed thicker metallic samples and quantitative stress evaluation using X-rays with higher photon energy, which are important next steps.

The drawback of using a rotating spiral slit is the long measurement time. The exposure time in the measurements with the spiral slit is almost 10 times (and the sample rotation speed was decreased by approximately 10 times) that in the measurements with a conical slit [29]. Nevertheless, uncolored white pixels without orientations appeared not only at grain boundaries but also in grains, which implies that the intensities of diffraction spots were insufficient. Noisy diffraction images due to low diffraction intensities give rise to experimental errors for the evaluation of N. The errors decrease the completeness of an orientation map. A solution for the drawback is to increase in the aperture ratio of the spiral slit. Although the four spiral-shaped inclined gaps were prepared in this demonstration, the aperture ratio can be increased by adding the number of the spiral-shaped inclined gaps. The decrease in stiffness of the tungsten blade plates due to additional spiral-shaped inclined through-grooves can be prevented because the blades are fixed to the rigid holder.

4. Conclusions

We herein proposed a scanning 3DXRD method with a rotating spiral slit. By sweeping Bragg angles using the rotating spiral slit, we can obtain diffraction images with patterns of diffraction spots from multiple grains in a gauge volume in a thick metallic sample independently of crystalline phases. From such diffraction images, orientation and stress maps can be reconstructed for various metals and alloys. Validation experiments were conducted with a 50-keV X-ray microbeam for two single-phase metals, low-carbon steel for

BCC, and pure Cu for FCC. As a result of orientation mapping with a pixel size of 1.2 × 1.2 μm, polygonal grain shapes and clear grain boundaries were successfully observed for both BCC and FCC metals. This implies that scanning 3DXRD with a rotating spiral slit is adaptable to various metals and alloys, multiphase alloys, and quantitative stress evaluation. Furthermore, if a higher-energy microbeam is available, scanning 3DXRD with a rotating spiral slit will be applicable to larger and thicker deformed metallic samples instead of current miniature test or wire-shaped samples.

Author Contributions: Y.H. and D.S. designed the research. K.F., K.O. and N.K. fabricated the rotating spiral slit. Y.H. performed the experiments. Y.H. and H.K. wrote the manuscript. All authors discussed the results and commented on the manuscript. All authors have read and agreed to the published version of the manuscript.

Funding: This research received no external funding.

Data Availability Statement: The data presented in this study are available on request from the corresponding author. The data are not publicly available due to the policy of the authors' affiliation (an incorporated company).

Acknowledgments: The synchrotron radiation experiments were performed at the BL33XU beamline of SPring-8 with the approval of the Japan Synchrotron Radiation Institute (JASRI) (Proposal No. 2019A7002.

Conflicts of Interest: The authors declare no conflict of interest.

References

1. Margulies, L.; Lorentzen, T.; Poulsen, H.F.; Leffers, T. Strain tensor development in a single grain in the bulk of a polycrystal under loading. *Acta Mater.* **2002**, *50*, 1771–1779. [CrossRef]
2. Aydıner, C.C.; Bernier, J.V.; Clausen, B.; Lienert, U.; Tomé, C.N.; Brown, D.W. Evolution of stress in individual grains and twins in a magnesium alloy aggregate. *Phys. Rev. B* **2009**, *80*, 024113. [CrossRef]
3. Oddershede, J.; Schmidt, S.; Poulsen, H.F.; Sørensen, H. O.; Wright, J.; Reimers, W. Determining grain resolved stresses in polycrystalline materials using three-dimensional X-ray diffraction. *J. Appl. Crystallogr.* **2010**, *43*, 539–549. [CrossRef]
4. Wang, L.; Lind, J.; Phukan, H.; Kenesei, P.; Park, J.-S.; Suter, R.M.; Beaudoin, A.J.; Bieler, T.R.. Mechanical twinning and detwinning in pure Ti during loading and unloading ? An in situ high-energy X-ray diffraction microscopy study. *Scr. Mater.* **2014**, *92*, 35–38. [CrossRef]
5. Shade, P.A.; Blank, B.; Schuren, J.C.; Turner, T.J.; Kenesei, P.; Goetze, K.; Suter, R.M.; Bernier, J.V.; Li, S.F.; Lind, J.; et al. A rotational and axial motion system load frame insert for in situ high energy X-ray studies. *Rev. Sci. Instrum.* **2015**, *86*, 093902. [CrossRef]
6. Schuren, J.C.; Shade, P.A.; Bernier, J.V.; Li, S.F.; Blank, B.; Lind, J.; Kenesei, P.; Lienert, U.; Suter, R.M.; Turner, T.J.; et al. New opportunities for quantitative tracking of polycrystal responses in three dimensions. *Curr. Opin. Solid State Mater. Sci.* **2015**, *19*, 235–244. [CrossRef]
7. Wong, S.L.; Obstalecki, M.; Miller, M.P.; Dawson, P.R. Stress and deformation heterogeneity in individual grains within polycrystals subjected to fully reversed cyclic loading. *J. Mech. Phys. Solids* **2015**, *79*, 157–185. [CrossRef]
8. Juul, N.Y.; Winther, G.; Dale, D.; Koker, M.K.A.; Shade, P.; Oddershede, J. Elastic interaction between twins during tensile deformation of austenitic stainless steel. *Scr. Mater.* **2016**, *120*, 1–4. [CrossRef]
9. Turner, T.; Shade, P.; Bernier, J.; Li, S.F.; Schuren, J.; Kenesei, P.; Suter, R.; Almer, J. Crystal plasticity model validation using combined high-energy diffraction microscopy data for a Ti-7Al specimen. *Metall. Mater. Trans. A Phys. Metall. Mater. Sci.* **2016**, *48*, 627–647. [CrossRef]
10. Renversade, L.; Borbély, A. Evaluation of grain-average stress tensor in a tensile-deformed Al–Mn polycrystal by high-energy X-ray diffraction. *J. Appl. Crystallogr.* **2017**, *50*, 1144–1157. [CrossRef]
11. Pagan, D.C.; Shade, P.A.; Barton, N.R.; Park, J.-S.; Kenesei, P.; Menasche, D.B.; Bernier, J.V. Modeling slip system strength evolution in Ti-7Al informed by in-situ grain stress measurements. *Acta Mater.* **2017**, *128*, 406–417. [CrossRef]
12. Abdolvand, H.; Wright, J.; Wilkinson, A.J. Strong grain neighbour effects in polycrystals. *Nat. Commun.* **2018**, *9*, 171. [CrossRef]
13. Guillen, D.P.; Pagan, D.C.; Getto, E.M.; Wharry, J.P. In situ tensile study of PM-HIP and wrought 316L stainless steel and Inconel 625 alloys with high energy diffraction microscopy. *Mater. Sci. Eng. A* **2018**, *738*, 380–388. [CrossRef]
14. Oddershede, J.; Camin, B.; Schmidt, S.; Mikkelsen, L.P.; Sørensen, H.O.; Lienert, U.; Poulsen, H.F.; Reimers, W. Measuring the stress field around an evolving crack in tensile deformed Mg AZ31 using three-dimensional X-ray diffraction. *Acta Mater.* **2012**, *60*, 3570–3580. [CrossRef]
15. Sedmák, P.; Pilch, J.; Heller, L.; Kopeček, J.; Wright, J.; Sedlák, P.; Frost, M.; Šittner, P. Grain-resolved analysis of localized deformation in nickel-titanium wire under tensile load. *Science* **2016**, *353*, 559–562. [CrossRef]

16. Naragani, D.; Sangid, M.D.; Shade, P.A.; Schuren, J.C.; Sharma, H.; Park, J.-S.; Kenesei, P.; Bernier, J.V.; Turner, T.J.; Parr, I. Investigation of fatigue crack initiation from a non-metallic inclusion via high energy X-ray diffraction microscopy. *Acta Mater.* **2017**, *137*, 71–84. [CrossRef]
17. Pagan, D.C.; Beaudoin, A.J. Utilizing a novel lattice orientation based stress characterization method to study stress fields of shear bands. *J. Mech. Phys. Solids.* **2019**, *128*, 105–116. [CrossRef]
18. Shen, Y.-F.; Liu, H.; Suter, R.M. Voxel-based strain tensors from near-field high energy diffraction microscopy. *Curr. Opin. Solid State Mater. Sci.* **2020**, *24*, 100852. [CrossRef]
19. Reischig, P.; Ludwig, W. Three-dimensional reconstruction of intragranular strain and orientation in polycrystals by near-field X-ray diffraction. *Curr. Opin. Solid State Mater. Sci.* **2020**, *24*, 100851. [CrossRef]
20. Naragani, D.; Shade, P.; Musinski, W.; Boyce, D.; Obstalecki, M.; Pagan, D.; Bernier, J.; Beaudoin, A. Interpretation of intragranular strain fields in high-energy synchrotron X-ray experiments via finite element simulations and analysis of incompatible deformation. *Mater. Des.* **2021**, *210*, 110053. [CrossRef]
21. Poulsen, H.F.; Cook, P.K.; Leemreize, H.; Pedersen, A.F.; Yildirim, C.; Kutsal, M.; Jakobsen, A.C.; Trujillo, J.X.; Ormstrup, J.; Detlefs, C. Reciprocal space mapping and strain scanning using X-ray diffraction microscopy. *J. Appl. Crystallogr.* **2018**, *51*, 1428–1436. [CrossRef]
22. Yildirim, C.; Jessop, C.; Ahlström, J.; Detlefs, C.; Zhang, Y. 3D mapping of orientation variation and local residual stress within individual grains of pearlitic steel using synchrotron dark field X-ray microscopy. *Scr. Mater.* **2021**, *197*, 113783. [CrossRef]
23. Yu, T.; Du, Y.; Fan, G.; Barabash, R.; Juul Jensen, D.; Zhang, Y. In situ synchrotron X-ray micro-diffraction investigation of elastic strains in laminated Ti-Al composites. *Metals* **2021**, *11*, 668. [CrossRef]
24. Zhang, Y.; Yu, T.; Xu, R.; Thorborg, J.; Liu, W.; Tischler, J.; Godfrey, A.; Juul Jensen, D. Local residual stresses and microstructure within recrystallizing grains in iron. *Mater Charact.* **2022**, *191*, 112113. [CrossRef]
25. Hayashi, Y.; Setoyama, D.; Seno, Y. Scanning three-dimensional X-ray diffraction microscopy with a high-energy microbeam at SPring-8. *Mater. Sci. Forum* **2017**, *905*, 157–164. [CrossRef]
26. Hektor, J.; Hall, S.; Henningsson, N.; Engqvist, J.; Ristinmaa, M.; Lenrick, F.; Wright, J. Scanning 3DXRD measurement of grain growth, stress, and formation of Cu_6Sn_5 around a tin whisker during heat treatment. *Materials* **2019**, *12*, 446. [CrossRef]
27. Li, W.; Sharma, H.; Peter, K.; Ravi, S.; Sehitoglu, H.; Bucsek, A. Resolving intragranular stress fields in plastically deformed titanium using point-focused high-energy diffraction microscopy. *J. Mater. Res.* **2023**, *38*, 165–178. [CrossRef]
28. Hayashi, Y.; Hirose, Y.; Seno, Y. Polycrystal orientation mapping using scanning three-dimensional X-ray diffraction microscopy. *J. Appl. Crystallogr.* **2015**, *48*, 1094–1101. [CrossRef]
29. Hayashi, Y.; Setoyama, D.; Hirose, Y.; Yoshida, T.; Kimura, H. Intragranular three-dimensional stress tensor fields in plastically deformed polycrystals. *Science* **2019**, *366*, 1492–1496. [CrossRef]
30. Henningsson, N.A.; Hall, S.A.; Wright, J.P.; Hektor, J. Reconstructing intragranular strain fields in polycrystalline materials from scanning 3DXRD data. *J. Appl. Crystallogr.* **2020**, *53*, 314–325. [CrossRef]
31. Henningsson, A.; Hendriks, J. Intragranular strain estimation in far-field scanning X-ray diffraction using a Gaussian process. *J. Appl. Crystallogr.* **2021**, *54*, 1057–1070. [CrossRef]
32. Kutsal, M.; Poulsen, H.F.; Winther, G.; Sørensen, H.O.; Detlefs, C. High-resolution 3D X-ray diffraction microscopy: 3D mapping of deformed metal microstructures. *J. Appl. Crystallogr.* **2022**, *55*, 1125–1138. [CrossRef]
33. Sergueeva, A.V.; Zhou, J.; Meacham, B.E.; Branagan, D.J. Gage length and sample size effect on measured properties during tensile testing. *Mater. Sci. Eng. A* **2009**, *526*, 79–83. [CrossRef]
34. Nielsen, S.F.; Wolf, A.; Poulsen, H.F.; Ohler, M.; Lienert, U.; Owen, R.A. A conical slit for three-dimensional XRD mapping. *J. Synchrotron. Rad.* **2000**, *7*, 103–109. [CrossRef]
35. Poulsen, H.F.; Margulies, L.; Schmidt, S.; Winther, G. Lattice rotations of individual bulk grains: Part I: 3D X-ray characterization. *Acta Mater.* **2003**, *51*, 3821–3830. [CrossRef]
36. Winther, G.; Margulies, L.; Schmidt, S.; Poulsen, H.F. Lattice rotations of individual bulk grains Part II: Correlation with initial orientation and model comparison. *Acta Mater.* **2004**, *52*, 2863-?2872. [CrossRef]
37. Suzuki, K.; Shobu, T.; Shiro, A.; Toyokawa, H. Evaluation of Internal Stresses Using Rotating-Slit and 2D Detector. *Mater. Sci. Forum* **2014**, *772*, 15–19. [CrossRef]
38. *JIS G 3141*; Cold-Reduced Carbon Steel Sheet and Strip. Japanese Standards Association: Tokyo, Japan, 2011.

quantum beam science

MDPI

Article

Stress Measurement of Stainless Steel Piping Welds by Complementary Use of High-Energy Synchrotron X-rays and Neutrons

Yasufumi Miura [1,*], Kenji Suzuki [2], Satoshi Morooka [3] and Takahisa Shobu [3]

1 Central Research Institute of Electric Power Industry, 2-6-1 Nagasaka, Yokosuka-shi 240-0196, Japan
2 Faculty of Education, Niigata University, Niigata 950-2181, Japan
3 Materials Science Research Center, Japan Atomic Energy Agency, 2-4 Shirakata, Tokai-mura, Naka-gun, Ibaraki 319-1195, Japan; shobu.takahisa@jaea.go.jp (T.S.)
* Correspondence: ymiura@criepi.denken.or.jp

Abstract: Probabilistic fracture mechanics (PFM) is increasingly recognized as a viable approach for evaluating the structural integrity of nuclear components, such as piping, primarily affected by stress corrosion cracking (SCC). PFM analysis requires several input parameters, among which welding residual stress is critically important due to its significant influence on SCC initiation and propagation. Recently, a novel technique involving a double-exposure method (DEM) utilizing synchrotron X-rays was introduced as an effective means for measuring welding residual stress with high spatial resolution. In this paper, we applied DEM to assess the residual stress of a plate specimen, which was extracted from a welded pipe through electrical discharge machining. Consequently, detailed stress maps under a plane stress state were generated. Additionally, the residual stress distributions in the welded pipe under a triaxial stress state were evaluated using neutron diffraction. Based on these findings, we proposed a methodology to acquire detailed stress maps of welded pipes by combining high-energy synchrotron X-rays and neutron diffraction.

Keywords: welding residual stress; austenitic stainless steel; synchrotron X-ray; neutron; double-exposure method

1. Introduction

Austenitic stainless steels, under the operational conditions of light water reactors (LWRs), are known to be prone to stress corrosion cracking (SCC). This phenomenon has been notably observed in the primary loop recirculation (PLR) piping of boiling water reactors (BWRs) [1]. One significant factor contributing to the initiation and propagation of SCC is the welding residual stress. Therefore, comprehending the characteristics of this stress is vital for assessing the structural integrity of LWR components susceptible to SCC [2]. In recent years, probabilistic fracture mechanics (PFM) has emerged as a significant method for evaluating the integrity of nuclear components prone to SCC. Several PFM codes specifically designed for piping applications have been developed [3–6]. PFM evaluations involving SCC require specific input parameters, such as SCC initiation time, crack growth rate, and residual stress distribution, within the assessment section. These parameters must be represented as probability distributions to ensure accurate evaluations. The absence of suitable probability distributions as input parameters could compromise the reliability of PFM evaluations. Since the initiation of SCC is stress-dependent, and the crack growth due to SCC relies on the stress intensity factor, the distribution of residual stress becomes a pivotal input parameter for evaluating SCC in piping using PFM codes. Considering that SCC in LWR components typically originates near welded areas, understanding the welding residual stress becomes particularly critical for accurate SCC evaluations. Hence, a thorough comprehension of the residual stresses in welded sections is imperative for effective SCC assessment.

X-ray diffraction methods are a prevalent nondestructive technique for measuring residual stresses [7]. Nonetheless, austenitic stainless steels, characterized by relatively large grain sizes and weld metals exhibiting anisotropy and coarser grains compared to the base metal, complicate X-ray stress measurements. Additionally, the X-ray penetration depth, limited to approximately 10 μm, restricts measurements to surface vicinities. This limitation poses significant challenges in assessing the welding residual stress distribution across the thickness of piping. Neutron diffraction methods [8], with their higher penetration depths, even in stainless steel, enable the measurement of internal stresses in structures like piping [9,10]. However, the difficulty in precisely collimating neutrons, coupled with the necessity of a certain measurement volume due to diffraction intensity constraints, means that unless a high-intensity neutron beam is employed, the spatial resolution is generally on the order of millimeters. This resolution may be inadequate for evaluating stress corrosion cracking (SCC) growth. Furthermore, large-diameter piping featuring relatively thick walls may still present challenges for neutron penetration depth. Consequently, residual stress distributions in welds are typically estimated using the finite element method (FEM), factoring in welding conditions rather than through direct measurements [11–13].

The deep hole drilling (DHD) method [14,15] measures stress distribution across the thickness of materials and is applicable to welded parts in plant components [10]. However, DHD is unable to measure depths within 0.5 mm from the surface. Moreover, the method's destructive nature precludes subsequent measurements near the initial measurement site, thereby limiting its utility for comprehensive evaluation of stress distribution in a welded joint.

The double-exposure method (DEM), utilizing high-energy synchrotron X-rays, was recently introduced as a nondestructive technique for measuring residual stress. Its effectiveness in coarse-grained metallic materials has been reported [16,17]. A notable advantage of the DEM is its elimination of the need to determine the diffraction center, a requirement that has previously limited the application of diffraction-based methods to coarse-grained structures and welds. This combination method of synchrotron radiation and neutrons shows promise for measuring welded metal parts, particularly those with coarse grains and anisotropy. Despite the use of synchrotron X-rays, direct measurement of welded pipes remains impractical due to the limited penetration depth in stainless steel. Consequently, it necessitates preparing thin plate specimens, approximately 5 mm in thickness, to facilitate X ray penetration. However, the specimen preparation process may induce stress release in the thickness direction of the plate, potentially altering the stress distribution from that in the original piping. Nevertheless, if the stress distribution prior to the stress release can be determined using neutron diffraction or similar methods, it becomes feasible to estimate the original residual stress distribution experimentally. This estimation can be achieved by combining pre-cutting stress information with stress data obtained post-cutting via DEM. Good agreement on the stress distribution between neutrons and synchrotron radiation was demonstrated in studies involving induction-hardened carbon steel [18]. Although the measurement did not focus on welds, the combined use of synchrotron radiation and neutrons could be a potent methodology.

In this study, we propose a method to create detailed residual stress maps in austenitic stainless steel piping welds. This method is based on the complementary application of high-energy synchrotron X-rays and neutrons.

2. Materials and Methods

2.1. Test Material

The test material was a welded joint in SUS316 piping. The base pipe was solution-annealed at 1060 °C after fabrication, and then water quenched. The joint featured a bevel angle of 60° and consisted of seven layers. The first layer was executed using gas tungsten arc welding (GTAW) with a stainless steel (JIS YS316L) insert ring. Layers 2 to

4 were fabricated using GTAW with stainless steel (JIS YS316L) welding wire. Layers 5 to 7 were constructed using shielded metal arc welding (SMAW) with a stainless steel (JIS ES316-16) welding rod, with the seventh layer being welded in two passes. Figure 1 presents a schematic diagram of the welded section. This combination of GTAW and SMAW represents one of the standard methodologies for welding medium-diameter stainless steel piping in nuclear fields. The heat input ranged from 10–20 kJ/cm for GTAW to 30–40 kJ/cm for SMAW. The excess weld metal and adjacent outer surface areas were smoothened through grinding and polishing. The base pipe had an outer diameter of 165.2 mm, a wall thickness of 14.3 mm, and a length of 200 mm. The thickness at the center of the weld line was approximately 15 mm, tapering to about 13 mm at positions 60 mm from the center due to the thinning process. The chemical compositions of the base metal and the weld materials are detailed in Table 1. A section near the weld line was removed using electrical discharge machining (EDM) to facilitate neutron diffraction measurements. Additionally, strain-free d_0 specimens, each 2 mm thick in the hoop direction of the pipe, and a plate specimen for the DEM measurements, 5 mm thick in the hoop direction, were also procured from this section via EDM. Figures 2 and 3 show a schematic diagram and a photograph of the test material, respectively.

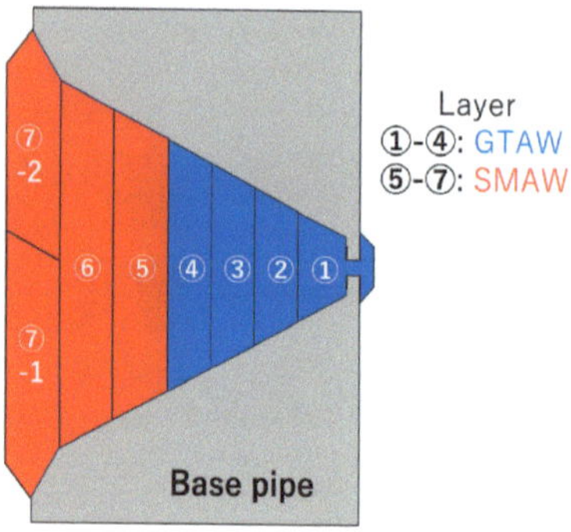

Figure 1. Schematic diagram of the weld part of the pipe used in this study.

Table 1. Chemical compositions of the base pipe and weld materials (mass%).

Element	C	Si	Mn	P	S	Ni	Cr	Mo	Cu	Fe
Base pipe	0.05	0.37	1.43	0.033	0.004	10.25	16.53	2.06	-	Bal.
Insert ring	0.012	0.36	1.78	0.023	0.001	12.09	19.44	2.36	0.27	Bal.
GTAW	0.017	0.41	1.88	0.005	0.002	11.38	19.61	2.31	0.01	Bal.
SMAW	0.055	0.41	1.40	0.029	0.009	12.08	19.22	2.33	0.26	Bal.

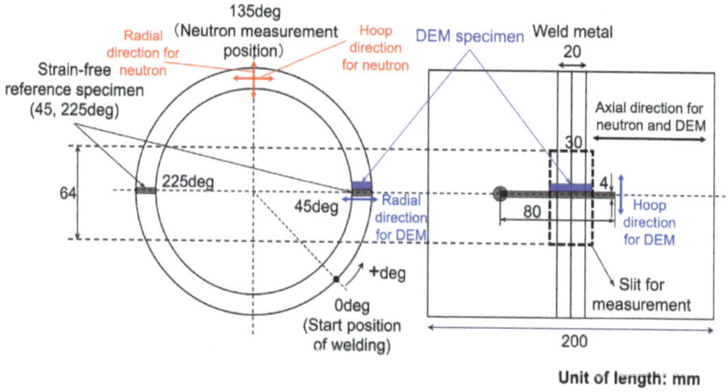

Figure 2. Schematic diagram of the welded pipe test material.

Figure 3. Photograph of the welded pipe test material (view from an angle of 225°).

2.2. Specimens

2.2.1. Strain-Free Reference Specimen

For the strain scanning method, it is necessary to prepare a specimen capable of assuming a strain-free state. Specimens with a thickness of 2 mm in the hoop direction of the pipe were extracted at angles of 45° and 225° utilizing wire-cut EDM. The elastic strain in these specimens was mitigated by cutting alternating slits from the inner and outer surfaces at 2 mm intervals using wire-cut EDM. These slits were then aligned and assembled to create a strain-free reference specimen with an overall thickness of 4 mm for subsequent measurements. The strain-free reference specimen is depicted in Figure 4. It is well-documented that weld metal in austenitic stainless steels typically contains a small fraction of δ-ferrite [19]. Microstructural analysis of a specimen obtained from the weld metal of the welded pipe confirmed the presence of approximately 8% δ-ferrite.

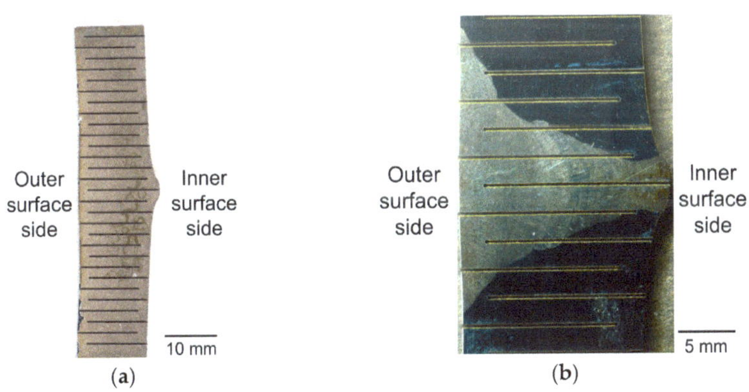

Figure 4. Strain-free reference specimen for neutron strain scanning: (**a**) photograph of the strain-free reference specimen, and (**b**) structure of the weld metal in the strain-free reference specimen.

2.2.2. DEM Specimen

The specimen for the double-exposure method (DEM) was prepared from a section at a 45° angle of the welded pipe specimen. This specimen was designed with dimensions of 5 mm in the hoop direction and 29.5 mm in the axial direction of the pipe. Its thickness in the radial direction was maintained equal to that of the pipe itself. Figure 5 features a photograph of the DEM specimen. In this image, the vertical center aligns with the center of the weld line, the left side represents the outer surface of the pipe, and the right side indicates the inner surface. Given that the thickness of this specimen was limited to 5 mm, it was presumed that the hoop stress within it would be zero.

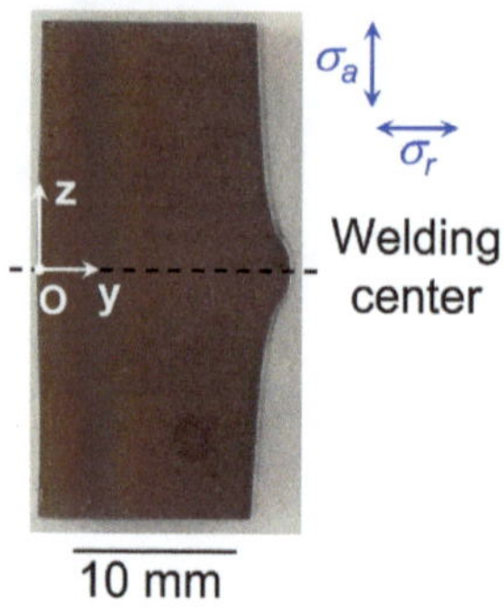

Figure 5. Photograph of the DEM specimen.

2.3. Neutron Stress Measurements

The thermal neutron flux emanating from Japan Research Reactor No. 3 (JRR-3) underwent monochromatization via a monochromator, culminating in extracting a neutron beam characterized by a wavelength of λ = 1.591 Å, tailored for these measurements. Stress assessments were conducted using the 311 diffractions from the γ-Fe and strain scanning methodology [8]. Data acquisition employed a one-dimensional 3He detector equipped with a 2 mm radial collimator on the detector side, facilitating the determination of the gauge volume. Diffraction patterns were derived by segregating and enumerating the diffracted neutrons through a 256-channel multichannel analyzer and subsequently approximated using a Gaussian function to determine the diffraction angle, 2θ. Photographs and schematic illustrations of the experimental setup for the neutron diffraction measurements are shown in Figure 6. These evaluations were primarily focused around a 135° angle, with specific measurement locations detailed in Figure 7. The strain was measured at each point in three orientations: axial, radial, and hoop. The slit dimensions were set at 2 mm width × 15 mm height for measurements in both axial and radial directions and 3 mm width × 3 mm height for the hoop direction. Consequently, the nominal gauge volume for measurements in the axial and radial directions assumed a prismatic shape elongated in the hoop direction. For the strain-free reference specimen, the slit dimensions were set to 2 mm width × 2 mm height in three orientations.

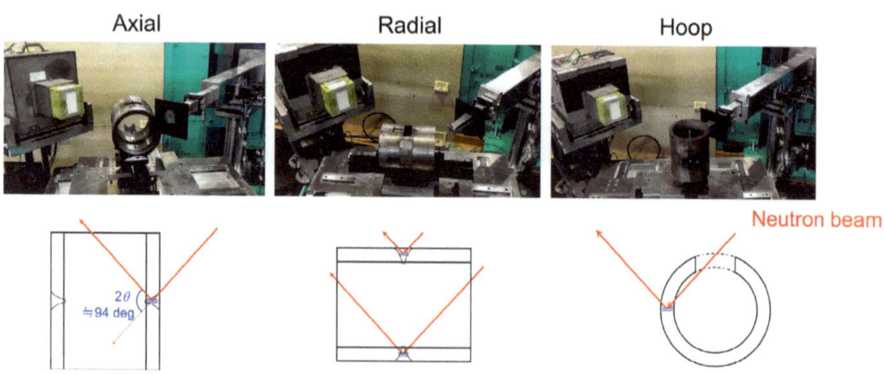

Figure 6. Photographs and schematic diagrams of the experimental configuration used for the neutron diffraction measurements.

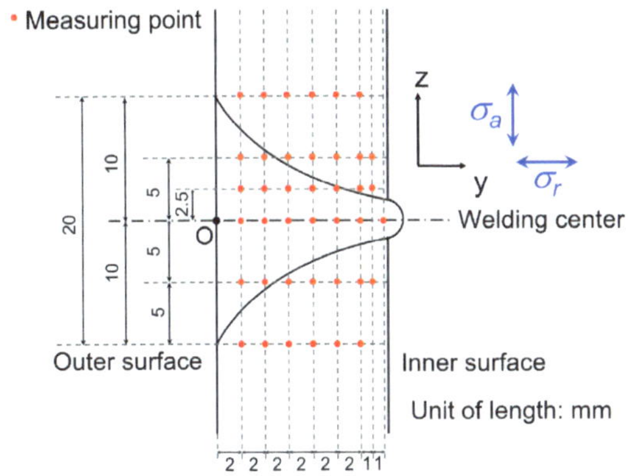

Figure 7. Measurement points for the neutron diffraction measurements.

The elastic strain, ε, at each measurement point was calculated from the following equation using the lattice spacing, d, at each measurement point and the strain-free lattice spacing, d_0, which were calculated using Bragg's law and the diffraction angle, 2θ:

$$\varepsilon = \frac{d - d_0}{d_0}. \quad (1)$$

The axial stress, σ_a, radial stress, σ_r, and hoop stress, σ_h, were obtained from the axial strain, ε_a, radial strain, ε_r, and hoop strain, ε_h, respectively, using the following equations:

$$\sigma_a = \frac{E}{1+v}\left[\varepsilon_a + \frac{v}{1-2v}(\varepsilon_a + \varepsilon_h + \varepsilon_r)\right], \quad (2)$$

$$\sigma_r = \frac{E}{1+v}\left[\varepsilon_r + \frac{v}{1-2v}(\varepsilon_a + \varepsilon_h + \varepsilon_r)\right], \quad (3)$$

$$\sigma_h = \frac{E}{1+v}\left[\varepsilon_h + \frac{v}{1-2v}(\varepsilon_a + \varepsilon_h + \varepsilon_r)\right]. \quad (4)$$

The diffraction elastic constant, E, and Poisson's ratio, ν, for the 311 lattice planes of SUS316 single crystals were determined to be 182.5 GPa and 0.307, respectively. These values were calculated using the Kröner model [20], incorporating elastic stiffness values of c_{11} = 206 GPa, c_{12} = 133 GPa, and c_{44} = 119 GPa [21]. The strain-free lattice spacing, d_0, was measured using a reference specimen; this measurement was conducted at $z = 0$ for the weld metal and at $z = -10$ for the base metal. Given that the variance in lattice spacing at each measurement point was relatively small and no significant directional dependence was observed, the average of all measured d_0 values was designated as d_0 for this study. The mean value of d_0 across all measurement points was 1.08471 Å, with a standard deviation of 1.3×10^{-4} Å. The stress range corresponding to this standard deviation, as calculated using the diffraction elastic constant, E, was found to be 23.7 MPa.

2.4. DEM with Synchrotron X-rays

Stress measurements were conducted using high-energy synchrotron radiation at the BL16XU beamline of SPring-8, employing an X-ray energy of 71.92 keV. The X-ray beam was shaped to 0.4 mm × 0.4 mm. The diffraction patterns were acquired using a PILATUS CdTe 300 K detector (Dectris Ltd., 5405 Baden-Daettwil, Switzerland), a two-dimensional detector installed on the 2θ arm of the diffractometer at BL16XU. The detector area was 83.8 mm × 106.5 mm, consisting of 487 × 619 pixels, each measuring 0.172 mm × 0.172 mm. The 2θ arm angle of the diffractometer was set to 9° to align with the diffraction angle of the 311 lattice planes at this X-ray energy. Measurements were performed in the axial and radial directions of the pipe, corresponding to the z and y axes, as shown in Figure 5.

Figure 8 presents a schematic diagram of the experimental setup during the DEM measurements. The X-ray beam passed through the DEM specimen, with diffracted X-rays detected vertically. Strain in the axial and radial directions of the pipe was measured by rotating the specimen 90° using the χ-cradle of the diffractometer. The detector positions labeled P1 and P2 were achieved by moving the detector on the 2θ arm by a distance of L, capturing diffraction patterns at both positions. The distance, L_0, between the specimen and P1 was 450 mm, while the distance, L, from P1 to P2 was 500 mm. Figure 9 displays a photograph of the experimental arrangement for the DEM measurements at detector position P1.

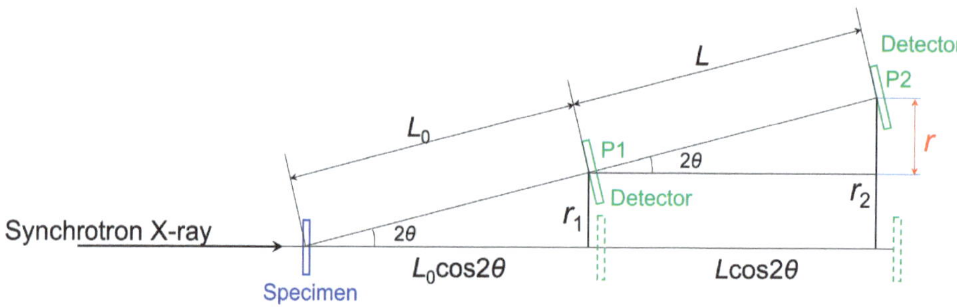

Figure 8. Schematic diagram of the experimental configuration used for the DEM measurements.

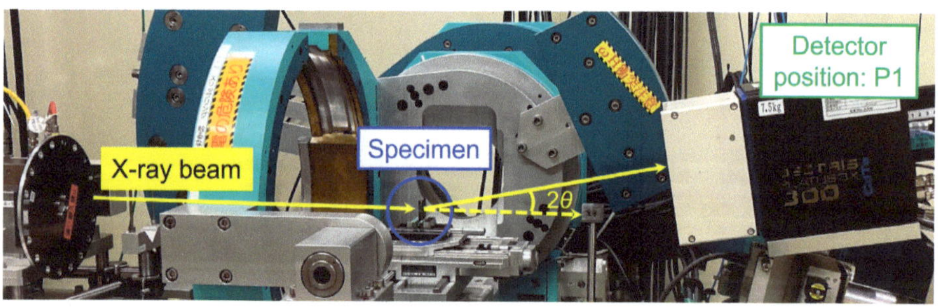

Figure 9. Photograph of the experimental setup used for the DEM measurements at detector position P1.

The diffraction radii, r_1 and r_2, were determined from the diffraction images measured at P1 and P2, respectively, and the diffraction radius, r, was calculated from the difference between r_2 and r_1, that is, $r = r_2 - r_1$. As shown in Figure 8, the diffraction angle, 2θ, can be expressed as follows:

$$2\theta = \sin^{-1}\left(\frac{r}{L}\right). \tag{5}$$

The effect of the diffracting crystal grains' positions within the specimen is canceled out by the definition of r, allowing for the acquisition of the diffraction angle, 2θ, without influence from the diffraction position. The coordinates for the DEM measurement are detailed in Table 2. While DEM analysis necessitates a stress-free reference lattice spacing, d_0, for the computation of elastic strain and stress, the lattice spacing deduced from the diffraction angle, 2θ, which in turn is derived from the diffraction pattern acquired by scanning from the inner to the outer surface at the $z = 14$ mm position, was defined as d_0. This spacing has been subsequently utilized in the strain calculation.

Table 2. DEM measurement coordinates (the y and z directions are indicated in Figure 5).

y Range (0.4 mm Pitch)	z Range (1.0 mm Pitch)
0.4~12.8	−7~−5, 6~10
0.4~14.0	−4~−2, 2~5
0.4~14.8	−1, ~1

3. Results and Discussion

3.1. Residual Stress Distribution Measured by Neutron Diffraction

A typical diffraction pattern and Gaussian fitting result of the neutron measurement are shown in Figure 10. In this study, we defined the residual stress error, $\Delta\sigma$, as the value calculated from Equations (2)–(4) using the residual strain error, $\Delta\varepsilon$, expressed as follows:

$$|\Delta\varepsilon| = \left|\frac{\Delta d}{d_0}\right|. \tag{6}$$

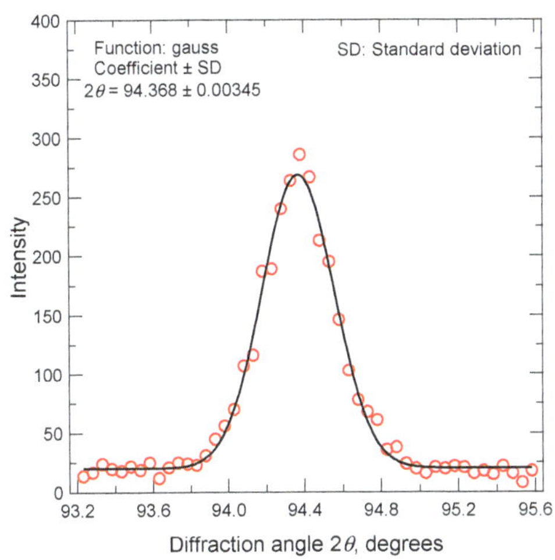

Figure 10. Diffraction profile in radial direction at the measurement point of y = 2, z = −10.

The residual strain error, $\Delta\varepsilon$, was determined from the lattice spacing error, Δd, which was itself derived from the standard deviation of the diffraction angle, $\Delta 2\theta$, ascertained through Gaussian fitting.

Figure 11 illustrates the stress distributions in the axial, radial, and hoop directions across the thickness of the welded pipe specimen. A consistent axial stress pattern was observed at all measurement locations, characterized by a decrease in stress from tension on the inner surface side towards the central area, followed by an increase towards the outer surface. In the weld heat-affected zone (HAZ), the axial stress on the inner surface side tended to diminish with increasing distance from the weld line, aligning with the reported FEM analysis results [11]. Figure 12 shows triaxial stress maps, which were developed from the data presented in Figure 11 and superimposed onto the DEM specimen. These maps were generated by designating each measurement point as a grid point and applying linear interpolation between these points. The maps revealed a relatively high tensile stress near the boundary between the weld metal and the HAZ on the inner surface of the pipe, correlating with the SCC initiation point observed in actual PLR piping of BWRs [1]. The stress distribution exhibited asymmetry relative to the weld line, potentially since the seventh layer of welding was performed using two-pass welding.

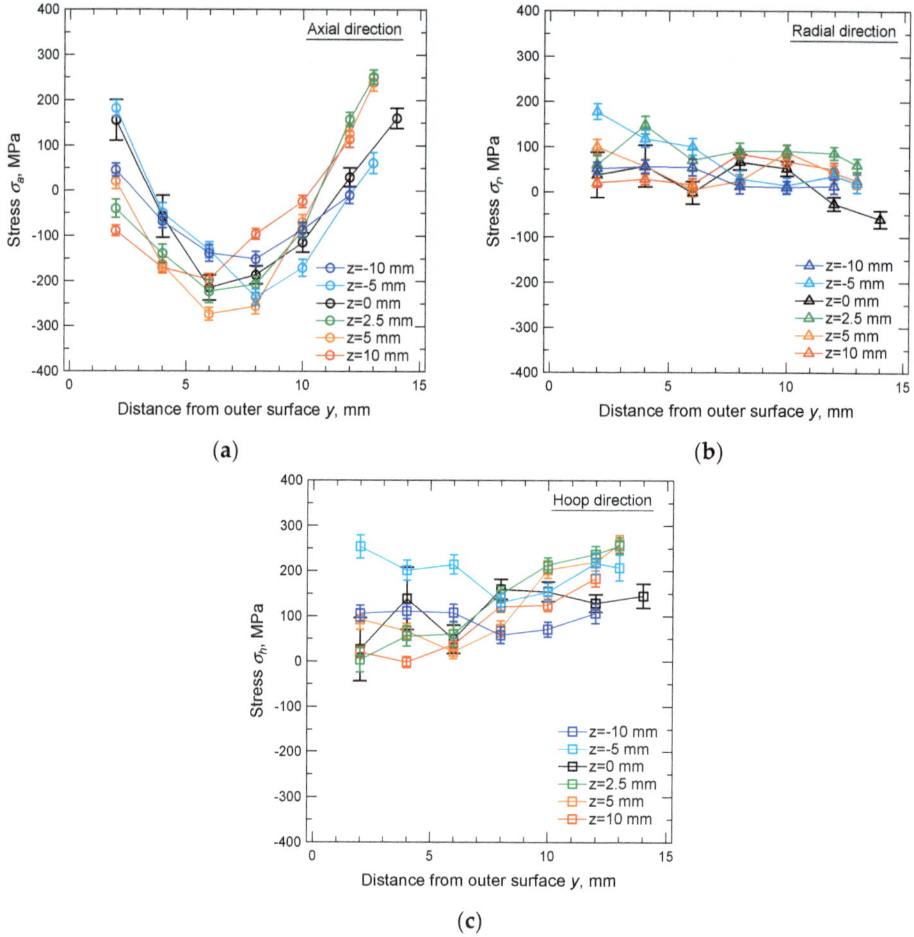

Figure 11. Stress distributions of the welded pipe specimen with respect to thickness: (**a**) axial direction, (**b**) radial direction, and (**c**) hoop direction.

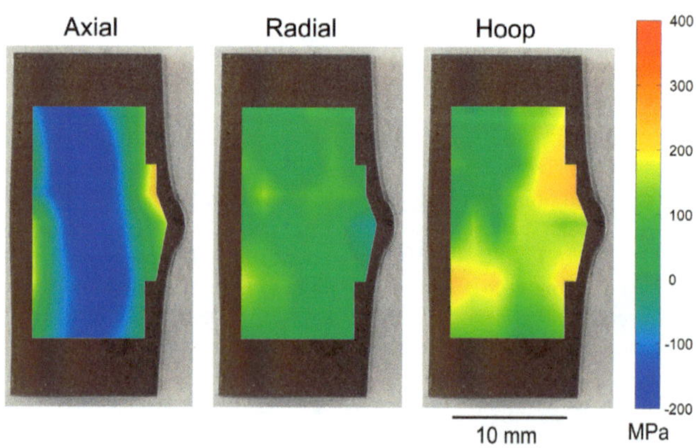

Figure 12. Welding residual stress maps of the welded pipe specimen measured by neutron diffraction.

3.2. 2D Stress Distributions Measured by DEM

Representative diffraction images from the specimen, acquired at detector positions P1 and P2, are illustrated in Figure 13. The bright areas in this figure denote the regions where integration was conducted to derive the diffraction curves. In these measurements, it is not possible to acquire a continuous diffraction ring due to the small beam size relative to the grain size. As depicted, the ring is discontinuous in both circumferential and radial directions, exhibiting non-uniform contrast. This discontinuity results from the presence of large dendrites and a limited number of crystal grains within the beam path. For most X-ray diffraction methods, accurately determining the diffraction angle, 2θ, from such a discontinuous ring presents a challenge. On the other hand, the similarity of the diffraction images within the integration areas of P1 and P2 is obvious. Given that the DEM technique leverages the difference between two diffraction images, it becomes feasible to ascertain the diffraction angle, 2θ, from these discontinuous images. Regarding the diffraction images captured at P1 and P2, the correlation between the diffraction radius and intensity was obtained through circumferential integration across a $\pm 5°$ range centered around the beam. This data was then approximated using a Gaussian function, enabling the calculation of the diffraction radii, r_1 and r_2, at P1 and P2, respectively. Subsequently, the diffraction angle, 2θ, was deduced using Equation (5).

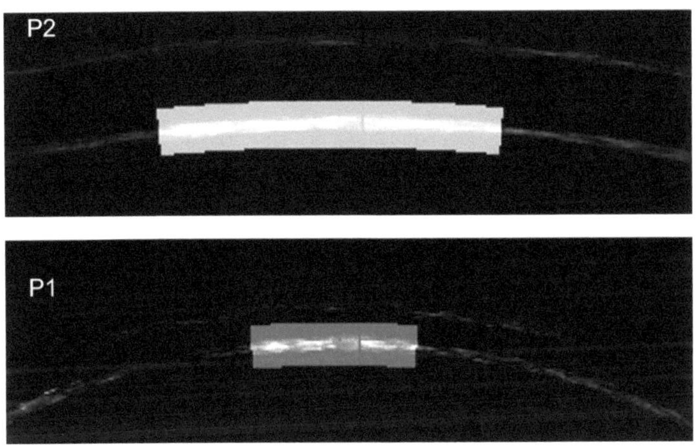

Figure 13. Representative diffraction images measured by the pixel detector at P1 and P2. The bright regions indicate the areas integrated to determine the diffraction angle, 2θ.

Since the strain obtained from Equations (1) and (5) is the value in the direction tilted by θ from the in-plane direction of the specimen, the strain in the in-plane direction was obtained using the following equations [17]:

$$\sigma_a = \frac{E}{(1+v)\cos^2\theta}\left[\varepsilon_a{}' + S'\left(\varepsilon_a{}' + \varepsilon_r{}'\right)\right], \tag{7}$$

$$\sigma_r = \frac{E}{(1+v)\cos^2\theta}\left[\varepsilon_r{}' + S'\left(\varepsilon_a{}' + \varepsilon_r{}'\right)\right], \tag{8}$$

$$S' = \frac{v}{(1+v)\cos^2\theta - 2v}, \tag{9}$$

where $\varepsilon_a{}'$ and $\varepsilon_r{}'$ are the axial and radial strains tilted by θ from the in-plane direction of the specimen, respectively, and E and ν are as defined above. The axial stress, σ_a, and radial stress, σ_r, were obtained using Equations (7)–(9) assuming a plane stress state. Residual stress maps of the DEM specimen created by linearly complementing the measured values as grid points are shown in Figure 14. In these measurements, there were some measurement points in the SMAW weld metal with coarse grains, where no diffraction image could be obtained within the circumferential integration region and strain measurement could not be performed. In such cases, the region without values around the measurement point is omitted.

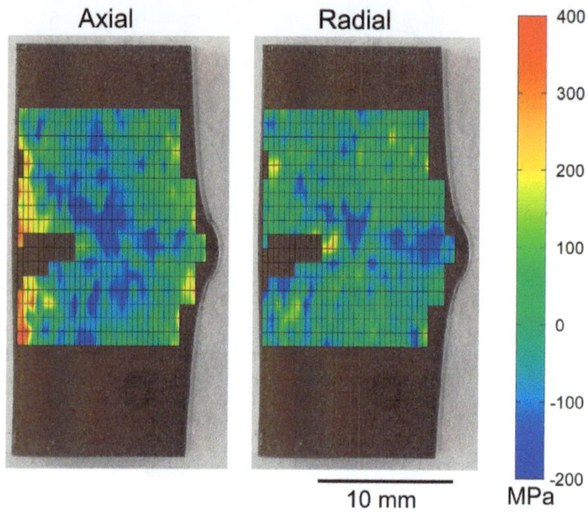

Figure 14. Residual stress maps of the DEM specimen under the plane stress condition.

In this measurement, the residual stress error was calculated from Δr and expressed as follows:

$$\Delta r = \sqrt{\Delta r_1{}^2 + \Delta r_2{}^2}. \tag{10}$$

Δr_1 and Δr_2 are the standard deviations of r_1 and r_2 determined by Gaussian fitting, respectively. The error range calculated using Equations (5)–(10) was about $\pm$15 MPa on average in both the axial and radial directions and about $\pm$40 MPa at points with low peak determination accuracy by Gaussian fitting.

Although the gauge volume and measurement location differ between the strain scanning method using neutrons and the DEM, the overall stress distributions were relatively similar. These distributions are characterized by axial stress decreasing from tension on the inner surface side toward the center and then increasing again towards the outer surface side. The DEM provided superior spatial resolution compared to neutron diffraction, enabling measurements close to the inner surface of the pipe weld. This capability is crucial for evaluating SCC initiation and propagation. However, the axial tensile stress observed on the inner surface of the pipe with DEM was not as high as that detected by neutron diffraction. This discrepancy is likely due to the relaxation of hoop stress during the slicing process in DEM specimen preparation.

3.3. *Evaluation of Triaxial Residual Stress by Combination of Neutron Diffraction and DEM*

To estimate the pre-relaxation stress distribution in the residual stress distribution maps measured by DEM, we utilize a method based on the hoop stress distribution obtained through neutron diffraction. This method involves approximating the hoop stress distribution, as presented in Figure 11, as a cubic function of the normalized distance from the outer surface, represented by y/t, where t denotes the pipe thickness at each 'z'

position. Under the assumption that the coefficients a, b, c, and d can be described as quartic functions of the distance z, we derive the following equation:

$$\begin{aligned}
\sigma_{h(y,\,z)} &= a\left(\tfrac{y}{t}\right)^3 + b\left(\tfrac{y}{t}\right)^2 + c\tfrac{y}{t} + d, \\
a &= 0.3341z^4 + 1.418z^3 - 34.07z^2 + 254.2z + 164.0, \\
b &= -0.9267z^4 - 2.462z^3 + 98.30z^2 + 429.7z - 537.9, \\
c &= 0.6901z^4 + 0.9443z^3 - 75.11z^2 - 167.7z + 589.1, \\
d &= -0.1530z^4 + 0.07210z^3 + 16.35z^2 - 4.224z - 42.68.
\end{aligned} \tag{11}$$

Next, we assume that the axial and radial strains of the DEM specimen obtained under the plane stress condition are subject to the hoop stress, $\sigma_{h(y,z)}$, in Equation (11) under the plane strain condition. The axial strain, ε_{a_DEM}, and radial strain, ε_{r_DEM}, obtained from Equations (7)–(9) assuming a plane stress state can be expressed by the following equations:

$$\varepsilon_{a\_DEM} = \frac{1}{E}(\sigma_{a\_DEM} - v\sigma_{r\_DEM}), \tag{12}$$

$$\varepsilon_{r\_DEM} = \frac{1}{E}(\sigma_{r\_DEM} - v\sigma_{a\_DEM}), \tag{13}$$

where E and ν are the same elastic constant and Poisson's ratio defined above. On the other hand, under the plane strain state, ε_a and ε_r can be expressed by the following equations:

$$\varepsilon_a = \frac{1}{E}\{\sigma_a - v(\sigma_r + \sigma_h)\}, \tag{14}$$

$$\varepsilon_r = \frac{1}{E}\{\sigma_r - v(\sigma_a + \sigma_h)\}. \tag{15}$$

By transforming Equations (14) and (15), the residual stresses under a triaxial stress state can be approximated by the following equations:

$$\sigma_a = \frac{E}{1 - v^2}(\varepsilon_{a\_DEM} + v\varepsilon_{r\_DEM}) + \frac{v}{1 - v}\sigma_{h(y,\,z)}, \tag{16}$$

$$\sigma_r = \frac{E}{1 - v^2}(\varepsilon_{r\_DEM} + v\varepsilon_{a\_DEM}) + \frac{v}{1 - v}\sigma_{h(y,\,z)}. \tag{17}$$

Under the assumptions of this paper, the stresses calculated by the equation may exceed the actual triaxial stresses. Nonetheless, considering the structural integrity assessment in relation to SCC, higher stress predictions yield conservative results and are thus deemed acceptable. Consequently, this method is validated as a simplified approach for triaxial stress evaluation in welded pipes.

Figure 15 shows residual stress maps, created similarly to Figure 14, utilizing stresses in three directions derived from Equations (11), (16), and (17). High axial stress around the HAZ near the inner surface and relatively high radial stress around the same HAZ were observed. These observations align with FEM analysis results reported in the literature [11]. Additionally, the residual stress maps suggest that SCC initiated at the HAZ tends to propagate towards the weld metal, a phenomenon corroborated by observations of the PLR piping of BWRs [1]. While extremely high tensile stresses near the outer surface are noticeable in Figure 15, these could be attributed to the grinding process during the welded pipe's fabrication. These results demonstrate that combining the DEM with neutron diffraction can uncover the residual stress distribution in the thickness direction with high spatial resolution. This includes the weld metal, which is typically challenging to measure using X-ray diffraction methods. The proposed method holds promise not only for measuring residual stresses in various welds but also for validating welding residual stress evaluation results obtained through numerical analysis, and such validations could also contribute to the advancement of numerical analysis methodologies.

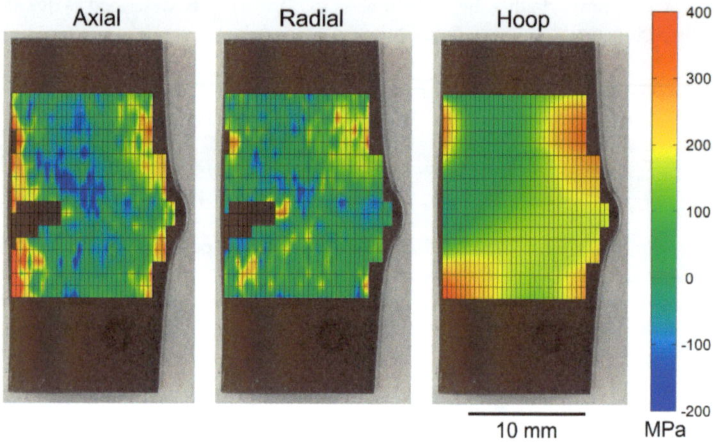

Figure 15. Modified residual stress maps of the DEM specimen considering the hoop stress measured by neutron diffraction.

4. Conclusions

In this study, we introduced a method for assessing the triaxial stress distribution in stainless steel piping welds by synergistically utilizing high-energy synchrotron X-ray radiation and neutron diffraction. The conclusions drawn from this research are as follows:

1. The DEM is an effective tool for stress evaluation in welds and coarse grains using high-energy synchrotron X-rays. Its effectiveness is primarily due to its low susceptibility to errors arising from the diffraction positions of the coarse grains in the weld.
2. Residual stresses in welded piping were quantified using the strain scanning method with neutron diffraction. The triaxial stress distribution, as determined by this method, was modeled using a polynomial equation, enabling the creation of a hoop stress map.
3. The axial and radial stresses, obtained from the DEM using high-energy synchrotron X-ray radiation, were integrated with the hoop stresses measured via neutron diffraction. This integration, based on a plane strain assumption, facilitated the development of a comprehensive residual stress map for the weld in a triaxial stress state.

Author Contributions: Conceptualization, K.S. and Y.M.; methodology, K.S.; software, K.S.; experiments, all authors; data analysis, Y.M.; writing—original draft preparation, Y.M. All authors have read and agreed to the published version of the manuscript.

Funding: This research received no external funding.

Data Availability Statement: Data are contained within the article.

Acknowledgments: The synchrotron radiation experiments at SPring-8 were performed with the approval of the Japan Synchrotron Radiation Research Institute (JASRI) Program Advisory Committee as follows: No. 2021A5051 and No. 2022A5051. The neutron experiments were carried out under the Inter-University Program for the Joint Use of JAEA/QST Facilities, Nuclear Professional School, School of Engineering, the University of Tokyo, at the Research Reactor JRR-3, JAEA (No. 22039).

Conflicts of Interest: The authors declare no conflict of interest.

References

1. Suzuki, S.; Takamori, K.; Kumagai, K.; Sakashita, A.; Yamashita, N.; Shitara, C.; Okumura, Y. Stress Corrosion Cracking in Low Carbon Stainless Steel Components in BWRs. *E-J. Adv. Maint.* **2009**, *1*, 1–29.

2. *JSME S NAI-2020*; Codes for Nuclear Power Generation Facilities—Rules on Fitness-for-Service for Nuclear Power Plants. Japan Society of Mechanical Engineers: Tokyo, Japan, 2020.
3. Rudland, D.; Harrington, D.; Dingreville, R. Development of the Low Probability of Rupture (xLPR) Version 2.0 Code. In Proceedings of the ASME Pressure Vessel and Piping Conference PVP2015, PVP2015-45134, Boston, MA, USA, 19–23 July 2015.
4. Homiack, M.; Facco, G.; Benson, M.; Erikcson, M.; Harrington, C. *Extremely Low Probability of Rupture Version 2 Probabilistic Fracture Mechanics Code*; NUREG-2247; U.S. Nuclear Regulatory Commission: Washington, DC, USA, 2021; p. 20555-0001.
5. Mano, A.; Yanaguchi, Y.; Katsuyama, J.; Li, Y. Improvement of Probabilistic Fracture Mechanics Analysis Code PASCAL-SP with Regards to PWSCC. *J. Nucl. Eng. Radiat. Sci.* **2019**, *5*, 31501. [CrossRef]
6. Machida, H.; Arakawa, M.; Yamashita, N.; Yoshimura, S. Development of Probabilistic Fracture Mechanics Analysis Code for Pipes with Stress Corrosion Cracks. *J. Power Energy Syst.* **2009**, *3*, 103–113. [CrossRef]
7. *JSMS-SD-5-02*; Standard Method for Stress Measurement—Steels. The Society of Materials Science: Kyoto, Japan, 2005.
8. Allen, A.J.; Hutchings, M.T.; Windsor, C.G.; Andreani, C. Neutron diffraction methods for study of residual stress fields. *Adv. Phys.* **1985**, *34*, 445–473. [CrossRef]
9. Suzuki, H.; Katsuyama, J.; Tobita, T.; Morii, Y. Residual stress measurement of large scaled welded pipe using neutron diffraction method—Effect of SCC crack propagation and repair weld on residual stress distribution. *Q. J. Jpn. Weld. Soc.* **2011**, *29*, 294–304. [CrossRef]
10. Bouchard, P.J.; George, D.; Santisteban, J.R.; Bruno, G.; Dutta, M.; Edwards, L.; Kingston, E.; Smith, D.J. Measurement of the stresses in stainless steel pipe girth weld containing long and short repair. *Int. J. Press. Vessel. Pip.* **2005**, *82*, 299–310. [CrossRef]
11. Machida, H. Influence of Non-Destructive Examination Performance on Reliability of Pipes Having Stress Corrosion Cracks. *Trans. JSME* **2011**, *77*, 1798–1813. [CrossRef]
12. Maekawa, A.; Serizawa, H.; Nakacho, K.; Murakawa, H. Fast Finite Element Analysis of Weld Residual Stress in Larg-Diameter Thick-Walled Stainless Steel Pipe Joints and Its Experimental Validation. *Q. J. Jpn. Weld. Soc.* **2013**, *31*, 129s–133s. [CrossRef]
13. Dai, P.; Wang, Y.; Li, S.; Lu, S.; Feng, G.; Deng, D. FEM analysis of residual stress induced by repair welding in SUS304 stainless steel pipe butt-welded joint. *J. Manuf. Process.* **2020**, *58*, 975–983. [CrossRef]
14. George, D.; Smith, D.J. Through thickness measurement of residual stresses in a stainless steel cylinder containing shallow and deep weld repairs. *Int. J. Press. Vessel. Pip.* **2005**, *82*, 279–287. [CrossRef]
15. Hilson, G.; Simandjuntak, S.; Flewitt, P.E.J.; Hallam, K.R.; Pavier, M.J.; Smith, D.J. Spatial variation of residual stresses in a welded pipe for high temperature applications. *Int. J. Press. Vessel. Pip.* **2009**, *86*, 748–756. [CrossRef]
16. Suzuki, K.; Shiro, A.; Toyokawa, H.; Saji, C.; Shobu, T. Double-Exposure Method with Synchrotron White X-ray for Stress Evaluation of Coarse-Grain Materials. *Quantum Beam Sci.* **2020**, *4*, 25. [CrossRef]
17. Suzuki, K.; Kura, K.; Miura, Y.; Shiro, A.; Toyokawa, H.; Kajiwara, K.; Shobu, T. A study on Stress Measurement if Weld Part using Double Exposure Method. *J. Soc. Mater. Sci. Jpn.* **2022**, *71*, 1005–1012. [CrossRef]
18. Holmberg, J.; Steuwer, A.; Stormvinter, A.; Kristoffersen, H.; Haakanen, M.; Berglund, J. Residual stress state in an induction hardened steel bar determined by synchrotron and neutron diffraction compared to results from lab-XRD. *Mater. Sci. Eng. A* **2016**, *667*, 199–207. [CrossRef]
19. Chopra, O.K. Effects of Thermal Aging on Fracture Toughness and Charpy-Impact Strength of Stainless Steel Pipe Welds. NUREG/CR-6428, Rev. 1; U.S. Nuclear Regulatory Commission: Washington, DC, USA, 2018; p. 20555-0001.
20. Kröner, E. Berechnung der elastischen Konstanten des Vierkristalls aus den Konstanten des Einkristalls. *Z. Physik* **1958**, *151*, 504–518. [CrossRef]
21. Suzuki, K. X-ray and Mechanical Elastic Constants for Cubic System by Kröner Model. Available online: https://x-ray.jsms.jp/kroner/kroner_c.html (accessed on 16 May 2023).

quantum beam science

MDPI

Article

Relationship between Internal Stress Distribution and Microstructure in a Suspension-Sprayed Thermal Barrier Coating with a Columnar Structure

Yasuhiro Yamazaki [1,*], Keisuke Shinomiya [2], Tadaharu Okumura [2], Kenji Suzuki [3], Takahisa Shobu [4] and Yuiga Nakamura [5]

1 Graduate School of Engineering, Chiba University, Chiba 263-8522, Japan
2 Graduate School of Science and Engineering, Chiba University, Chiba 263-8522, Japan
3 Faculty of Education, Niigata University, Niigata 950-2181, Japan
4 Research Center for Materials Science, Nuclear Science Research Division, Japan Atomic Energy Agency, Sayo 679-5148, Japan
5 Diffraction and Scattering Promotion Office, Japan Synchrotron Radiation Research Institute, Sayo 679-5148, Japan
* Correspondence: y.yamazaki@chiba-u.jp

Abstract: The suspension plasma spray (SPS) method is expected to become a novel coating method because it can achieve various microstructures using a suspension with submicron spray particles. Thermal barrier coatings (TBCs) with a columnar structure, which might achieve high strain tolerance, can be obtained using the SPS technique. This study evaluated the internal stress distribution of the suspension-plasma-sprayed thermal barrier coating (SPS-TBC) with different columnar structures using hybrid measurement using high-energy synchrotron X-ray diffraction analysis and laboratory low-energy X-rays. The relationship between the microstructure and the internal stress distribution of the SPS-TBC was discussed on the basis of the experimental results. In addition, the in-plane internal stress was decreased by decreasing the column diameter. The thin columnar microstructure of the SPS-TBC has superior strain tolerance. The internal stresses in the SPS-TBC are periodic decrements caused by stress relaxation in porous layers in its column.

Keywords: suspension plasma spray; thermal barrier coating; columnar structure; internal stress distribution; effect of microstructure; high-energy synchrotron X-ray diffraction analysis

1. Introduction

Thermal barrier coatings (TBCs) have been widely applied to blades and vanes in hot section components of land-based gas turbines and aero engines because the turbine inlet temperature of the operating gas reaches 1600 °C or higher to achieve high efficiency [1–4]. The typical TBCs are composed of the yttria-stabilized zirconia type ceramic top coat to insulate against heat conduction and an oxidation-resistant metallic bond coating containing Co, Ni, Cr, Al, Y, etc. In general, the atmospheric plasma spray (APS) technique has been used as a coating method for the TBC top coat of land-based gas turbines. During the APS process, the spray powders with tens of micrometers are deposited onto the bond coat surface as melted or half-melted by the plasma flame of a spray gun. The deposited lamellar microstructure of APS-TBCs with many inter-splat pores has a good heat-shielding effect. However, because APS-TBC has relatively low strain tolerance, its thermal cycle fatigue resistance is inferior to the electric beam physical vapor-deposited (EB-PVD) TBC with a columnar structure. Many studies have been conducted for the durability [5–7] and mechanical properties [8–10] of APS-TBCs and EB-PVD TBCs. APS-TBCs have a lamellar microstructure, including pores and low thermal conductivity [11]. On the other hand, EB-PVD TBCs have a columnar structure with excellent thermal stress relaxation proper-ties; however, they are not suitable for large components because of their coating process in vacuums [12].

The importance of renewable energy systems, represented by solar and wind power generation systems, has recently increased to achieve a low carbon emission society. However, the supply capacity of such renewable energy systems is sensitive to the season's conditions and weather. For example, the short-time power fluctuation in the solar power generation system reaches 25% of total capacity. Hence, other power generation systems are needed as back-up suppliers. The land-based gas turbines are excellent as a supply–demand adjustment generation system to absorb the fluctuation of renewable energy systems because of their superb bootability and high flexibility [13]. However, the hot section components in supply–demand adjustment gas turbine systems, such as turbine blades and vane, are exposed to more severe service conditions. In such cases, TBCs must have low thermal conductivity as an APS-TBC and excellent thermal shock resistance as an EB-PVD TBC. The suspension plasma spray (SPS) technique is expected to be a novel coating method that can produce various microstructures, such as dense, porous, columnar coatings, etc., because of its ability to control a coated microstructure using a suspension of submicron-sized fine powder [14–23]. Many recent efforts have been made to research microstructure control [16–18], thermal properties [18,19], damage morphologies [17,19,21], etc., in the suspension-plasma-sprayed thermal barrier coatings (SPS-TBCs). The authors experimentally demonstrated that TBCs with a cauliflower-like columnar microstructure deposited by the SPS technique exhibit high thermal cycle fatigue resistance compared with the conventional APS-TBC [24]. In addition, they revealed the effect of the microstructure of the SPS-TBC with a cauliflower-like columnar micro-structure on thermal cycle fatigue lives, that is, the fine columnar structure has higher resistance to thermal cycle fatigue [25].

Turbine blades and vanes with TBC are subjected to complex and cyclic internal stress during the operation because of the thermal coefficient mismatch between the top coat as well as the substrate and thermally grown oxide generated at the interface between the top coat and bond coat; consequently, the thermal cycle damage, such as the delamination, and cracking of the top coat, occurs [25,26]. Therefore, internal stress measurements have been conducted in the TBC system using X-rays [9,27,28]. One of the authors developed a hybrid method by using high-energy X-rays from synchrotron radiation of which the penetration depth from the surface is more profound than 10 μm and low-energy X-rays with a shallow penetration of less than 1 μm and measured the internal stress distribution in APS-TBCs and EB-PVD TBCs by using the hybrid method [29–31]. Knipe et al. investigated the in situ measurement of strain response in EB-PVD TBC with a columnar structure under the thermal gradient using a high-energy synchrotron X-ray [32]. However, no investigation has been conducted to evaluate the internal stress distribution in SPS-TBCs with a cauliflower-like microstructure.

This study evaluated the internal stress distribution of the SPS-TBC with different columnar structures using hybrid measurement using high-energy synchrotron X-ray diffraction analysis and laboratory low-energy X-rays. Furthermore, the relationship between the microstructure and the internal stress distribution of the SPS-TBC was discussed on the basis of the experimental results.

2. Experimental Procedures

2.1. Materials and Specimen Preparation

In this study, eight weight present yttria-stabilized zirconia (8YSZ) were used as the ceramic thermal insulate top coat. Three types of specimens with a ceramic top coat containing columns with different diameters were prepared using the suspension plasma spray technique. Each type of specimen was denoted as SPS-F, SPS-B, and SPS-C specimens, respectively. Zhou et al. reported that the spray surface condition (the metallic bond coat surface) could control the column structure of the SPS-TBC [33]. In this study, however, the suspension feedstock supply rate during the suspension plasma spray process was controlled to change the microstructure of the samples; the suspension feedstock rate was increased in the order of SPS-C > SPS-B > SPS-F (the bond coat surface in each specimen was in the same condition, and the stand-off distance was 70 mm in fixed).

The conventional polycrystalline Ni-base superalloy IN738LC was utilized as the substrate material. The preparation method of substrate specimens was as follows. First, the disk specimens with 1.2 mm thickness were cut from a round bar of IN738LC by wire electric discharge machining. After the electrical cutting process, the machining heat-affected zone on the spray surface of the substrate specimen was removed by mechanical polishing. The final geometry of the substrate specimen was 1 mm in thickness and 20 mm in diameter. The CoNiCrAlY bond coat was sprayed on the substrate disk using the high-velocity oxide fuel (HVOF) technique with JP-5000 as a spray torch. The bond coat thickness was approximately 100 μm. The suspension with fine 8YSZ spray particles with sub-micrometer diameters was sprayed on the bond coat surface using a plasma spray torch (100HE). The top coat of all type specimens had a thickness of approximately 200 μm. The spray conditions of the top coat and bond coat are listed in Table 1.

Table 1. Spray conditions of the bond coat and top coat.

	Bond Coat	Top Coat
Spray method	HVOF	SPS
Spray material	CoNiCrAlY	Ethanol-based YSZ suspension (solid-phase content in suspension was 25 wt.%)
Spray torch	JP-5000	100HE
Plasma gas	N/A	Ar + N2 + H2 (flow rate:199 SLPM)
Plasma power	N/A	105 kW

The cross-sectional microstructures of the specimens are shown in Figure 1. The top coat of each specimen has a columnar structure with different column diameters. The laminated structure consistent with dense and porous layers can be observed in the column (the details are shown later). The column diameter of each specimen, d, was represented by their diameter at the half TC thickness from the results of SEM observations. The column diameter, d, of the SPS-F, SPS-B, and SPS-C specimens were approximately 100, 120, and 180 μm, respectively.

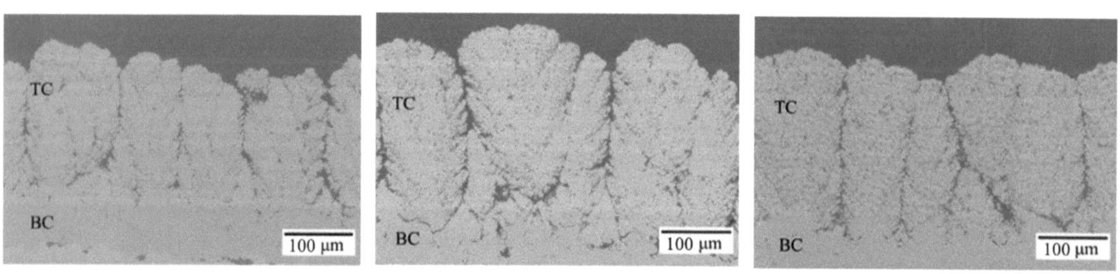

(a) (b) (c)

Figure 1. Microstructures of the top coat in SPS-TBCs: (**a**) SPS-F specimen; (**b**) SPS-B specimen; (**c**) SPS-C specimen [24]; TC: top coat, BC: bond coat.

All specimens were pre-thermally exposed before the X-ray analysis using a muffle furnace at 1000 °C for 300 h. The pre-thermal exposure aimed to generate a TGO layer at the interface between the top coat and the bond coat and to apply the in-plane biaxial tensile loading in the top coat. In this study, the in-plane biaxial tensile loading in the top coat was applied by using the deformation of the substrate caused by the pre-thermal exposure. Figure 2 shows the results of the substrate surface profile of the specimen (after removing the top coat by polishing) measured from the cross-sectional image before and after the pre-thermal exposure. As shown in Figure 2, the specimen was deformed convexly after the pre-thermal exposure. Consequently, the in-plane tensile load in the top coat was applied. By using elastic finite analysis, the in-plane strain at the top-coat/bond-coat interface caused by substrate deformation was estimated to be 0.12% in tension. Therefore, substrate deformation during the pre-thermal exposure occurred because of the release of residual stress in the substrate and bond coat, which was induced by bond coat spraying and shot blasting on the substrate surface [34].

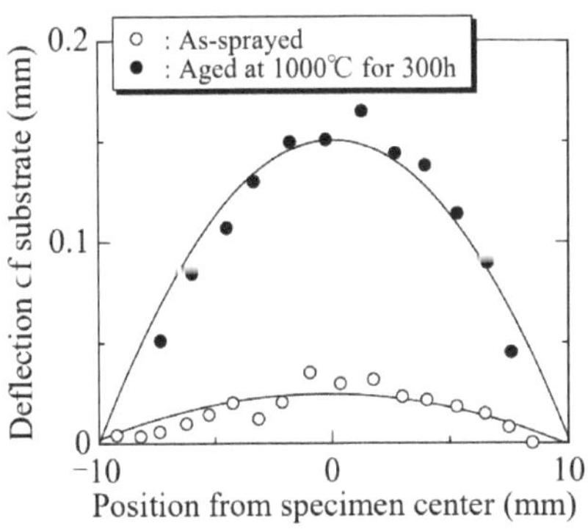

Figure 2. Geometry change of substrate by thermal aging [34].

After pre-thermal exposure, the top coat surface was polished by using diamond paste with a diameter of 9, 3, and 1 μm to prepare the top coat thickness to 100 μm. In this study, the original TC thickness after the suspension plasma spray process, in other words, the distance between the original TC surface and the TC/BC interface, was measured from the difference in the thickness of each specimen before and after the TC spray. The TC thickness of each specimen was controlled by the removed thickness by polishing from the original TC surface before the X-ray analysis.

2.2. Hybrid X-ray Deflection Analysis to Measure the Internal Stress in SPS-TBC

This study evaluated the internal stress distribution in the top coat of SPS-TBC using the hybrid method using high-energy synchrotron X-ray diffraction analysis and laboratory low-energy X-rays [35]. The difference between high- and low-energy X-rays is their penetration depth from the sample surface due to their energy difference. The former has a penetration depth more profound than 10 μm, whereas on the other hand, the latter less than 1 μm. The detail of the measurement method is described below.

Considering that the equi-biaxial stress sat in the in-plane direction, it can be assumed for the top coat of TBCs, the relationship between the stress components and the measured diffraction angle θ is represented as follows:

$$2\theta = 2\theta_0 - \frac{2(1+\nu_x)}{E_x}(\sigma_{in} - \sigma_{out}) \tan\theta_0 \sin^2\psi - \frac{2}{E_x}\sigma_{out}\tan\theta_0 + \frac{4\nu_x}{E_x}\sigma_{in}\tan\theta_0 \quad (1)$$

where θ_0 is the diffraction angle for a strain-free condition; σ_{in} and σ_{out} are the stress components for the in-plane and out-of-plane direction, respectively. E_x and ν_x are Young's modulus and Poisson's ratio for X-ray analysis, respectively.

Considering that the low-energy X-rays for general X-ray stress measurement used in a laboratory have a shallow penetration depth, the stress evaluated by using low-energy X-rays is limited to the stress state at the sample surface. In addition, the evaluation area is in the plane stress condition ($\sigma_{out} \approx 0$). The following equation can be obtained by partial differentiation using $\sin^2 \psi$, which is shown in Equation (1):

$$\frac{\partial 2\theta}{\partial \sin^2 \psi} = -\frac{2(1+\nu_x)}{E_x}\sigma_{in} \tan \theta_0 \tag{2}$$

Hence, the in-plane stress, σ_{in}, can be obtained from the gradient of the $2\theta - \sin^2 \psi$ curve. The stress evaluated with low-energy X-rays, $\sigma_{X\text{-}ray}$, is equal to the in-plane stress, which is calculated as follows:

$$\sigma_{X\text{-}ray} = \sigma_{in} \tag{3}$$

In this study, the distribution of the in-plane stress, σ_{in}, was obtained by the sequential polishing method [29], in which σ_{in} is measured by sequentially reducing the specimen thickness by polishing. In contrast, the stress state is far from the plane stress condition because of its deeper penetration depth when high-energy synchrotron X-rays are used for the stress analysis. Thus, the gradient of the $2\theta - sin^2\psi$ diagram measured by high-energy synchrotron X-rays is calculated as follows:

$$\frac{\partial 2\theta}{\partial \sin^2 \psi} = -\frac{2(1+\nu_x)}{E_x}(\sigma_{in} - \sigma_{out}) \tan \theta_0 \tag{4}$$

Thus, the stress measured by high-energy synchrotron X-rays, σ_{syn}, is measured as follows:

$$\sigma_{syn} = (\sigma_{in} - \sigma_{out}) \tag{5}$$

The X-ray penetration depth is varied with its incident angle; however, the effect of it on the experimental result is not huge then, negligible for simplicity in the following analysis.

The internal stress distribution of $(\sigma_{in} - \sigma_{out})$ was obtained by using the $sin^2\psi$ method with a side-inclination configuration [29–31]. Internal stress measurements were conducted using high-energy synchrotron X-rays with a four-circle goniometer at the beamline BL02B1 in Japan Synchrotron Research Institute, SPring-8. The measurement conditions are summarized in Table 2. Figure 3 shows the diffraction profile obtained by high-energy synchrotron X-rays. The diffractions from ZrO_2 (422) and (224) were used, considering the X-ray penetration depth and reflection intensity. The scintillator (Ohyo Koken Kogyo Co., Ltd., Tokyo, Japan, SP-10) was used to measure the intensity of the X-ray diffraction profile in this study. The used scintillator consisted of the scintillator substance and the photomultiplier tube. The single transparent crystal of thallium-activated sodium iodine, Nal(Tl), was used as the scintillator substance. When diffracted radiation reaches the scintillator, it is absorbed by the Nal(Tl) scintillator and converted into fluorescent light. The converted fluorescent light is converted and amplified by the photomultiplier tube into an electrical pulse for counting. In this study, a long solar slit was installed at the front of the scintillation counter to suppress the broadening of the diffraction profile.

Table 2. Conditions of stress measurement by synchrotron high-energy X-ray.

Beamline	BL02B1
Wave length	0.01736 nm (71.4 KeV)
Size of slit (V × H)	0.2 mm × 1.0 mm
Analysis method	Constant penetration depth method (Side-inclination configuration)
Crystal	ZrO_2
Diffraction plane	(422) + (224)
$2\theta_0$	9.4992°
Stress constant [28]	−11,492 MPa/deg
$\sin^2 \psi$	0–0.7

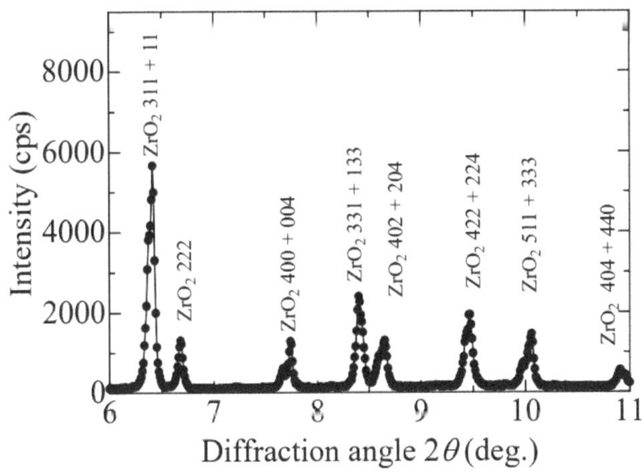

Figure 3. Typical diffraction profile obtained by high-energy synchrotron X-rays.

On the other hand, the in-plane stress in the ceramic top coat, σ_{in}, was obtained by using low-energy X-rays in the laboratory. The low-energy X-ray measurement was conducted using the cos α method with Pulstec Industries μ-X360s25. Table 3 shows the measurement conditions. In this study, the internal stress distribution in the TC was evaluated by repeating the X-ray stress measurements after polishing the TC surface to reduce the TC thickness.

Table 3. Conditions of stress measurement by low-energy X-ray.

Analysis method	cos α
Characteristic X-ray	Cr-Kα
Tube voltage	30 V
Tube current	1.2 mA
Crystal	ZrO_2
Diffraction plane	111
$2\theta_0$	15.2578°
Stress constant	−193,160 MPa/deg
Young's modulus [36]	107 GPa
$\sin^2\psi$	0–0.7

3. Results and Discussion

Solid symbols in Figure 4 represent typical diffraction profiles obtained by the $\sin^2\psi$ method with a side-inclination configuration using high-energy X-rays. As shown in Figure 4, the profiles of the (422) and (224) diffraction planes of ZrO_2 overlap within the selected 2θ diffraction angle range between 9.4° and 9.6°. The profiles of the (422) and (224) diffraction planes were approximated by Gaussian functions in the present study, and the superimposed (422) and (224) diffraction profiles expressed by the following equation were fitted with the least-square approximation to the profiles measured by high-energy X-rays.

$$A_{422}exp\left\{-\frac{(\theta-\theta_{422})}{B_{422}}\right\}+A_{224}exp\left\{-\frac{(\theta-\theta_{224})}{B_{224}}\right\} \tag{6}$$

where, θ_{422} and θ_{224} are the peak diffraction angles for the profiles of (422) and (224) diffraction planes, respectively, and A_{422}, A_{224}, B_{422}, B_{224} are the fitting constants. The first and second terms in Equation (6) correspond to the (422) and (244) diffraction profiles, respectively. Then, the peak diffraction angles for the (422) and (224) diffraction planes, θ_{422} and θ_{224}, respectively, were obtained from the approximated profiles. Figure 5 shows the typical relationship between the peak diffraction angles ($2\theta_{422}$ and $2\theta_{224}$) and $\sin^2\psi$ obtained from the approximate curves. σ_{syn} ($= \sigma_{in} - \sigma_{out}$) was obtained from the $2\theta_{422} - \sin^2\psi$ and $2\theta_{224} - \sin^2\psi$ relations using Equations (4) and (5).

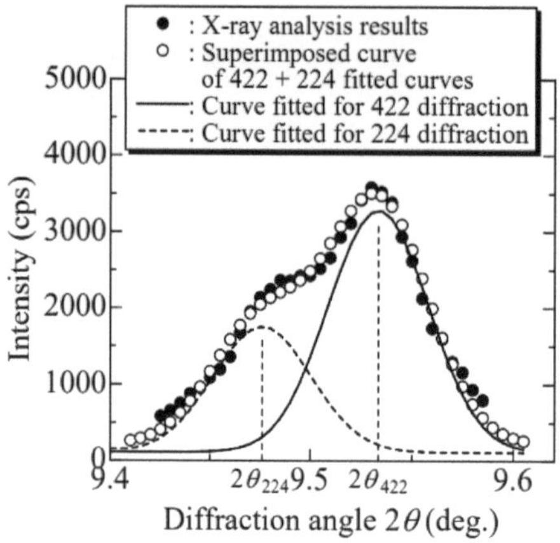

Figure 4. Typical diffraction profile used to evaluate σ_{syn}; SPS-B specimen [34].

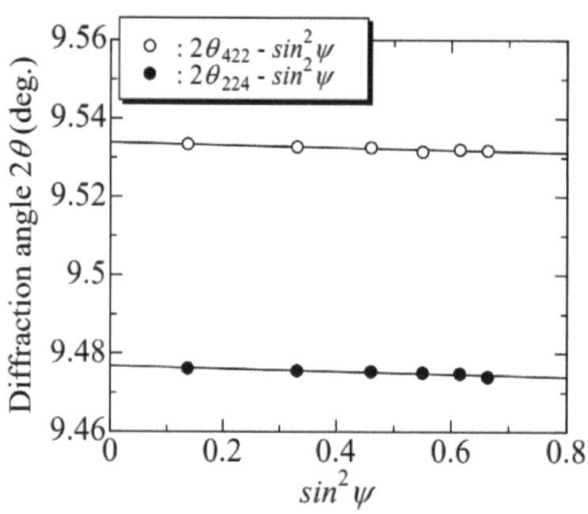

Figure 5. Typical $2\theta - \sin^2\psi$ diagram used to evaluate σ_{syn}; SPS-B specimen [34].

Figure 6 shows the internal stress distribution of σ_{syn} for each specimen. The horizontal axis in Figure 6 exhibits the position of the measurement point from the TC/BC interface. For high-energy X-ray analysis, the distance in Figure 6 was calculated from the difference in the thickness of TC (= distance between the TC/BC interface and the TC surface) and the penetration depth of the X-ray. On the other hand, in the case of low-energy analysis, the horizontal axis means the TC thickness after sequential polishing. As shown in Figure 6, the stress distribution of $\sigma_{X\text{-}ray}$ evaluated with low-energy X-rays is also represented. In addition, no differences in σ_{syn} in each specimen obtained from profiles of the (422) and (224) diffraction planes of ZrO_2 were observed. Moreover, the σ_{syn} for each specimen is decreased periodically. The periodic decrement in the σ_{syn} distribution will be discussed later.

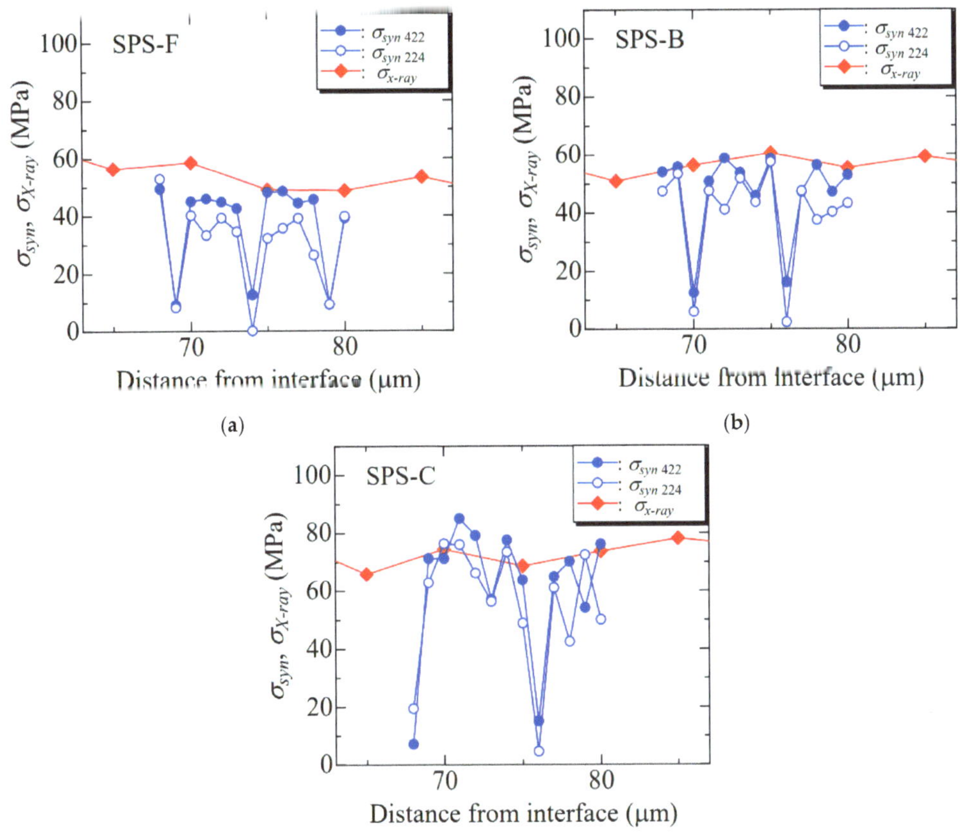

Figure 6. Internal stress distribution evaluated by high- and low-energy X-rays: (**a**) SPS-F specimen; (**b**) SPS-B specimen; (**c**) SPS-C specimen.

The distribution of σ_{syn} evaluated by high-energy synchrotron X-rays was compared with $\sigma_{X\text{-}ray}$ evaluated with low-energy X-rays. Although σ_{syn} in each specimen was almost comparable to $\sigma_{X\text{-}ray}$ except for the presence or absence of the periodic decrease in stress distributions, as shown in Figure 6. The results in Figure 6 indicate that σ_{syn} was almost equal to the in-plane stress component, σ_{in}, and no out-of-plane stress component was applied ($\sigma_{out} \approx 0$), not only at the surface of the top coat but also inside the top coat (at the positions from the TC/BC interface ranging between 65 and 80 μm).

Figure 7 shows the influence of the microstructure of the ceramic top coat on the internal stress distribution obtained from experiments using high-energy synchrotron X-rays. As shown in Figure 7, σ_{syn} was the average value evaluated from profiles of the (422) and (224) diffraction planes of ZrO_2. σ_{syn} decreased with decreasing the column diameter of the top coat. The results in Figure 7 indicate that the thinner the column microstructure of the SPS-TBC, the higher the strain tolerance. Our previous investigation about the influence of the microstructure on the thermal cycle fatigue strength in the SPS-TBC revealed that the thermal cycle fatigue life increased with decreasing the column diameter [24]. Based on the results shown in Figure 7, the superior thermal cycle fatigue strength of the thin column microstructure of the SPS-TBC was achieved by high strain tolerance. However, it was reported that, in an EB-PVD TBC, the residual internal stress decreased with increasing the column diameter [29]. The microstructure of EB-PVD TBC was controlled by the rotation rate of the substrate in the EB-PVD process. The column diameter of the EB-PVD TBC increased with increasing the rotation rate of the substrate. In addition, the porosity in the column also increased with increasing the rotation rate and affected the residual internal stress. The effect of the pore in the column of SPS-TBC will be discussed later.

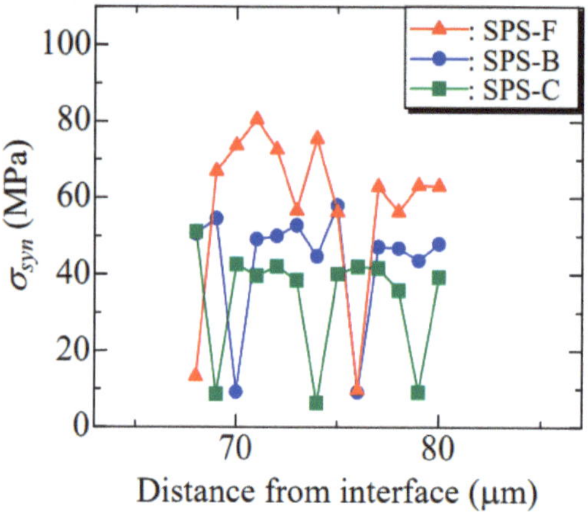

Figure 7. Effect of microstructure on internal stress distribution.

As shown in Figure 6, the periodical decrement in the internal stress distribution of σ_{syn} was observed, although it was not in the results obtained by low-energy X-rays. On the other hand, porous layers were periodically in the SPS-TBC column (Figure 8). Figure 9 shows the periods of the low internal stress layer and the porous one as a function of the column diameter. As shown in Figure 9, both periods were almost identical, indicating that the internal stress was released at the porous layer. In addition, the porous layer period was proportionate to the column diameter. The internal stress of the EB-PVD TBC with a columnar structure was reduced with increasing the porosity [29]. It was not easy to measure the porosity of the tested SPS-TBCs from the cross-sectional SEM images. However, it seems from Figure 1 that the area fraction of pores in the column increased with increasing the column diameter. Therefore, not only the volume fraction of pores but also the density of the periodic porous layers in the column, in other words, the period of the porous layer, might affect the reduction of internal stress in the SPS-TBC. Based on these results, the higher thermal fatigue strength of the SPS-F specimen might also be related to the higher density of the porous layer, which can reduce internal stress.

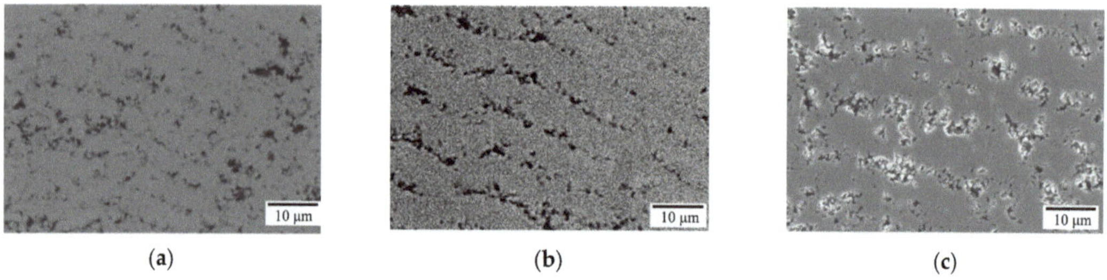

Figure 8. Typical microstructure in the column of the SPS-TBCs: (**a**) SPS-F specimen; (**b**) SPS-B specimen; (**c**) SPS-C specimen.

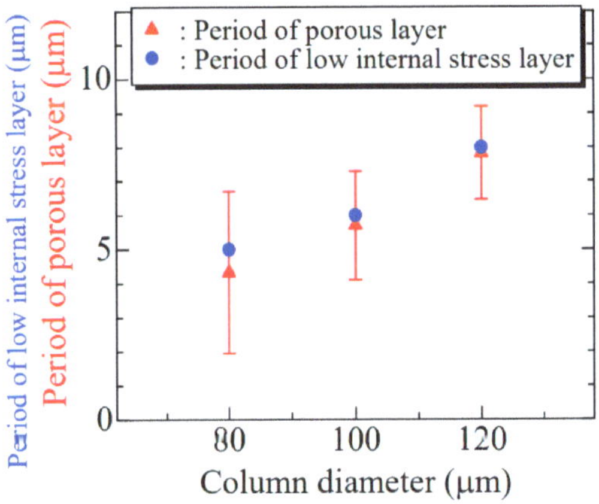

Figure 9. Periods of the low internal stress layer and the porous one as a function of the column diameter.

As shown in Figure 6, the periodic stress decrement cannot be observed in the stress distribution evaluated by low-energy X-rays. As mentioned above, the low-energy X-rays for general X-ray stress measurement used in a laboratory have a shallow penetration depth. Therefore, the stress evaluated using low-energy X-rays is limited to the stress state at the sample surface. Considering that the porous layer at the specimen surface was detached during polishing, the stress decrement layer cannot be evaluated using low-energy X-rays. These results indicate that the internal stress distribution, including the periodic decrement, can be only evaluated using high-energy synchrotron X-rays.

4. Summary Remarks

This study investigated the internal stress of the SPS-TBC with different columnar structures using hybrid measurements, including high-energy synchrotron X-ray diffraction analysis and laboratory low-energy X-rays. Stress distributions ranging between 65 and 80 μm from the top-coat/bond-coat interface were obtained when in-plane biaxial tensile loading was applied in the top coat.

The following conclusions were drawn:

(1) In the top coat of the SPS-TBC, the internal stress was almost in the in-plane stress state, and the out-of-plane stress component was practically equal to zero;
(2) The in-plane internal stress in the top coat decreased by decreasing the column diameter. The thin column microstructure of the SPS-TBC had superior strain tolerance;
(3) In measuring internal stress using high-energy synchrotron X-rays, periodic stress decrements were observed in the internal stress distribution because internal stress was released at the porous layer.

Author Contributions: Y.Y.: conceptualization, methodology, data curation, writing—original draft, writing—review and editing, visualization, supervision, project administration, funding acquisition. K.S. (Keisuke Shinomiya): investigation, data curation, visualization. T.O.: investigation, data curation, visualization. K.S. (Kenji Suzuki): methodology, investigation. T.S.: methodology, investigation. Y.N.: methodology, investigation. All authors have read and agreed to the published version of the manuscript.

Funding: This research was funded by the Japan Society for the Promotion of Science (JSPS) KAKENHI Grant Number 21K03768.

Data Availability Statement: Data will be made available on request.

Acknowledgments: The first author appreciated the financial support from the Japan Society for the Promotion of Science (JSPS) KAKENHI Grant Number 21K03768. The synchrotron radiation experiments were carried out at SPring-8 with approval of the Japan Synchrotron Radiation Research Institute (Proposal No. 2020A1605, 2021B1716 and 2022B0554). The authors are obliged to T. Hamada, Y. Habu, K. Takaki (TOCALO Co., Ltd., Hyogo, Japan), and H. Suzuki (Pulstec Industrial Co., Ltd., Shizuoka, Japan) for their assistance in the experiments.

Conflicts of Interest: The authors declare no conflict of interest.

References

1. Padture, N.P.; Gell, M.; Jordan, E.H. Thermal barrier coatings for gas-turbine engine applications. *Science* **2002**, *296*, 280–284. [CrossRef]
2. Vaßen, R.; Kaßner, H.; Stuke, A.; Hauler, F.; Hathiramani, D.; Stöver, D. Advanced thermal spray technologies for applications in energy systems. *Surf. Coat. Technol.* **2008**, *202*, 4432–4437. [CrossRef]
3. Bakan, E.; Vasßen, R. Ceramic top coats of plasma-sprayed thermal barrier coatings: Materials, Processes, and Properties. *J. Therm. Spray Technol.* **2017**, *26*, 992–1010.
4. Li, G.-R.; Yang, G.-J.; Chen, X.-F.; Li, C.-X.; Li, C.-J. Strain/sintering co-induced multiscale structural changes in plasma-sprayed thermal barrier coatings. *Ceram. Int.* **2018**, *44*, 14408–14416. [CrossRef]
5. Yamazaki, Y.; Schmidt, A.; Scholz, A. The determination of delamination resistance in thermal barrier coating system by four-point bending tests. *Surf. Coat. Technol.* **2006**, *201*, 744–754. [CrossRef]
6. Trunova, O.; Beck, T.; Herzog, R.; Steinbrech, R.W.; Singheiser, L. Damage mechanisms and lifetime behavior of plasma sprayed thermal barrier coating systems for gas turbines—Part I: Experiments. *Surf. Coat. Technol.* **2008**, *202*, 5027–5032. [CrossRef]
7. Yamazaki, Y.; Yoshida, T.; Fukanuma, H.; Ohno, N. Thermal Cycle Damage and Residual Adhesion Strength in a Thermal Barrier Coating. *J. Soc. Mater. Sci. Jpn.* **2009**, *58*, 168–174. [CrossRef]
8. Yamazaki, Y.; Kinebchi, T.; Fukanuma, H.; Ohno, N. Influence of the spray process parameters and the thermal exposure on the mechanical properties of the free-standing air-plasma sprayed thermal barrier coating. *J. Jpn. Therm. Spray Soc.* **2007**, *44*, 88–95.
9. Yamazaki, Y. Correlation between the mechanical properties and splat microstructures of an air plasma sprayed thermal barrier coatings. *J. Solid Mech. Mater. Eng.* **2008**, *2*, 1275–1286. [CrossRef]
10. Okazaki, M.; Yamagishi, S.; Yamazaki, Y.; Ogawa, K.; Waki, H.; Arai, M. Adhesion Strength of Ceramic Top Coat in Thermal Barrier Coatings Subjected to Thermal Cycles: -Effects of Thermal Cycle Testing Method and Environment. *Int. J. Fatigue* **2013**, *53*, 33–39. [CrossRef]
11. Arikawa, H.; Kojima, Y. Heat Resistant Coatings for Gas Turbine Materials. *J. Surf. Finish. Soc. Jpn.* **2001**, *52*, 11–15.
12. Matsumoto, K.; Kawagishi, K.; Harada, H. Development of Thermal Barrier Coating Technology. *Mater. Jpn.* **2013**, *52*, 469–474. [CrossRef]
13. Watanabe, K.; Takahashi, T. Development of Gas Turbine Based on Energy Planning of Japan and Issues on GTCC Contributing to Stability of Electric Power System. *J. Gas Turbine Soc. Jpn.* **2019**, *47*, 2–7.
14. Schlegel, N.; Ebert, S.; Mauer, G.; Vaßen, R. Columnar-structured Mg-Al-spinel thermal barrier coatings (TBCs) by suspension plasma spraying (SPS). *J. Therm. Spray Technol.* **2015**, *24*, 144–151. [CrossRef]
15. Schlegel, N.; Sebold, D.; Sohn, Y.; Mauer, G.; Vaßen, R. Cycling performance of a columnar-structured complex perovskite in a temperature gradient test. *J. Therm. Spray Technol.* **2015**, *24*, 1205–1212.
16. Ganvir, A.; Curry, N.; Björklund, S.; Markocsan, N.; Nylén, P. Characterization of microstructure and thermal properties of YSZ Coatings obtained by axial suspension plasma spraying (ASPS). *J. Therm. Spray Technol.* **2015**, *24*, 1195–1204. [CrossRef]
17. Guignard, A.; Maur, G.; Vaßen, R.; Stöver, D. Deposition and characteristics of submicronmeter-structured thermal barrier coatings by suspension plasma spraying. *J. Therm. Spray Technol.* **2012**, *21*, 416–424. [CrossRef]
18. Ganvir, A.; Curry, N.; Markocsan, N.; Nylén, P.; Joshi, S.; Vilemova, M.; Pala, Z. Influence of microstructure on thermal properties of axial suspension plasma-sprayed YSZ thermal barrier coatings. *J. Therm. Spray Technol.* **2016**, *25*, 202–212. [CrossRef]
19. Curry, N.; VanEvery, K.; Snyder, T.; Markocsan, N. Thermal Conductivity Analysis and Lifetime Testing of Suspension Plasma-Sprayed Thermal Barrier Coatings. *Coatings* **2014**, *4*, 630–650.
20. Xu, P.; Meng, G.; Liu, G.; Cpyle, T.; Pershin, L.; Mostaghimi, J. Columnar-structured thermal barrier coatings deposited via the water-based suspension plasma spray process. *J. Phys. D Appl. Phys.* **2022**, *55*, 204001. [CrossRef]
21. Curry, N.; Tang, Z.; Markocsan, N.; Nylén, P. Influence of bond coat surface roughness on the structure of axial suspension plasma spray thermal barrier coatings—Thermal and lifetime performance. *Surf. Coat. Technol.* **2015**, *268*, 15–23. [CrossRef]
22. Curry, N.; Mahade, S.; Venkat, A.; Joshi, S. Erosion performance of suspension plasma spray thermal barrier coatings—A comparison with state of art coatings. *Surf. Coat. Technol.* **2022**, *437*, 12811. [CrossRef]
23. Ma, X.; Ruggiero, P. Practical aspects of suspension plasma spray for thermal barrier coatings on potential gas turbine components. *J. Therm. Spray Technol.* **2018**, *27*, 591–602. [CrossRef]
24. Yamazaki, Y.; Shinomiya, K.; Hamaguchi, T.; Habu, Y.; Takagi, K. Effect of the microstructure of suspension plasma-sprayed thermal barrier coatings on their thermal cycling damage. *Surf. Coat. Technol.* **2023**, *456*, 129269. [CrossRef]

25. Wu, R.T.; Osawa, M.; Yokokawa, T.; Kawagishi, K.; Harada, H. Degradation Mechanisms of an Advanced Jet Engine Service-Retired TBC Component. *J. Solid Mech. Mater. Eng.* **2010**, *4*, 119–130. [CrossRef]
26. Ito, A.; Kobayashi, D. Degradation evaluation of 1300 °C-class 1st-stage gas turbine blades after long-term service. In Proceedings of the International Gas Turbine Congress 2015, Tokyo, Japan, 15–20 November 2015; pp. 614–619.
27. Matejicek, J.; Sampath, S. Intrinsic residual stresses in single splats produced by thermal spray processes. *Acta Mater.* **2001**, *49*, 1993–1999. [CrossRef]
28. Chun, L.; Zhang, X.; Chen, Y.; Carr, J.; Jacques, S.; Behnsen, J.; Michiel, M.; Xiao, P.; Cernik, R. Understanding the residual stress distribution through the thickness of atmosphere plasma sprayed (APS) thermal barrier coatings (TBCs) by high energy synchrotron XRD; digital image correlation (DIC) and image based modelling. *Acta Mater.* **2017**, *132*, 1–12.
29. Suzuki, K.; Shobu, T.; Wada, K.; Matsubara, H.; Kawamura, M. Microstructure and Residual Stress of EB-PVD TBCs Grown under Substrate Rotation. *J. Soc. Mater. Sci. Jpn.* **2008**, *57*, 674–680. [CrossRef]
30. Suzuki, K.; Matsumoto, K.; Kubo, T.; Machiya, S.; Tanaka, K.; Akiniwa, Y. Analysis on Residual Stress in Electron Beam-Physical Vapor Deposited Thermal Barrier Coating using Hard Synchrotron X-rays. *Trans. Jpn. Soc. Mech. Eng. Ser. A* **2005**, *71*, 1523–1529. [CrossRef]
31. Suzuki, K.; Kawamura, M.; Shobu, T.; Kubo, T.; Tanaka, K.; Akiniwa, Y. Residual stress and deformation characteristics of thermal barrier coatings on curved substrate. *J. Soc. Mater. Sci. Jpn.* **2006**, *55*, 634–640. [CrossRef]
32. Knipe, L.; Manero, A.; Siddiqui, S.; Meid, C.; Wischek, J.; Okasinski, J.; Almer, J.; Karlsson, A.M.; Bartsch, M.; Raghavan, S. Strain response of thermal barrier coatings captured under extreme engine environments through synchrotron X-ray diffraction. *Nat. Commun.* **2014**, *5*, 4559. [CrossRef]
33. Zhou, D.; Guillon, O.; Vassen, R. Development of YSZ thermal barrier coatings using axial suspension plasma spraying. *Coatings* **2017**, *7*, 120. [CrossRef]
34. Yamazaki, Y.; Shinomiya, K.; Okumura, T.; Suzuki, K.; Shobu, T.; Nakamura, Y. Internal stress distribution in suspension plasma sprayed thermal barrier coating subjected to in-plane tensile loading. *J. Jpn. Therm. Spray Soc.* **2023**, *60*. *In press*.
35. Suzuki, K.; Tanaka, K. Applications on High-Energy X-Rays to Stress Measurements of Thermal Barrier Coatings. *Textures Microstruct.* **2003**, *35*, 207–217. [CrossRef]
36. Yamazaki, Y.; Morikawa, M.; Hamaguchi, T.; Habu, Y.; Ohide, Y.; Takagi, K. Relationship between the mechanical properties and structure of a suspension plasma-sprayed thermal barrier coating with columnar microstructure. *Surf. Coat. Technol.* **2022**, *439*, 128430. [CrossRef]

quantum beam science

MDPI

Article

Accuracy of Measuring Rebar Strain in Concrete Using a Diffractometer for Residual Stress Analysis

Ayumu Yasue [1], Mayu Kawakami [2], Kensuke Kobayashi [1], Junho Kim [2], Yuji Miyazu [2], Yuhei Nishio [3], Tomohisa Mukai [4], Satoshi Morooka [5] and Manabu Kanematsu [2,*]

1 Graduate School of Science and Technology, Tokyo University of Science, 2641 Yamazaki, Noda 278-8510, Chiba, Japan
2 Faculty of Science and Technology, Tokyo University of Science, 2641 Yamazaki, Noda 278-8510, Chiba, Japan
3 Department of Fire Engineering, Building Research Institute, 1 Tachihara, Tsukuba 305-0802, Ibaraki, Japan
4 Building Department, National Institute for Land and Infrastructure Management, 1 Tachihara, Tsukuba 305-0802, Ibaraki, Japan
5 Materials Sciences Research Center, Japan Atomic Energy Agency, 2-4 Shirakata, Tokai, Naka 319-1195, Ibaraki, Japan
* Correspondence: manabu@rs.tus.ac.jp

Abstract: Neutron diffraction is a noncontact method that can measure the rebar strain inside concrete. In this method, rebar strain and stress are calculated using the diffraction profile of neutrons irradiated during a specific time period. In general, measurement accuracy improves with the length of the measurement time. However, in previous studies, the measurement time was determined empirically, which makes the accuracy and reliability of the measurement results unclear. In this study, the relationship between the measurement time and the measurement standard deviation was examined for reinforced concrete specimens under different conditions. The aim was to clarify the accuracy of the measurement of rebar stress using the neutron diffraction method. It was found that if the optical setup of the neutron diffractometer and the conditions of the specimen are the same, there is a unique relationship between the diffraction intensity and the rebar stress standard deviation. Furthermore, using this unique relationship, this paper proposes a method for determining the measurement time from the allowable accuracy of the rebar stress, which ensures the accuracy of the neutron diffraction method.

Keywords: reinforced concrete; rebar stress; neutron diffraction method; non-destructive test; bond; accuracy intensity; measurement time; standard deviation

1. Introduction

The local bond behavior between rebar and concrete can be evaluated by measuring the stress in the rebar inside the concrete. In previous studies [1–3], bond behavior was evaluated by applying strain gauges to the rebars inside the concrete and using them to measure the stress distribution of the rebars. Although the strain gauge method is characterized by its extremely high measurement accuracy, there is concern that applying the strain gauge to the rebar and handling the lead wire itself may affect the bond behavior. In addition, it is challenging to evaluate the stresses inside the cross-section of the rebar because the strain gauge can only measure the stress on the surface of the rebar, which is where the strain gauge is attached.

In contrast, the neutron diffraction method has attracted attention as a measurement method that can evaluate the rebar strain inside concrete using non-destructive and non-contact methods [4]. Neutrons also have excellent permeability through concrete and steel, which makes it possible to determine the stresses inside the rebar cross section. For these reasons, the neutron diffraction method has been applied to evaluate the strain distribution in bond tests of common [5–7], corroded [6], and hot-dip galvanized rebars [8]. In particular, previous studies [4,5] have reported that rebar stress distributions measured by neutron diffraction and strain gauge methods give different results, indicating the effectiveness of the neutron diffraction method in evaluating bond behavior.

However, very few studies have been conducted on the neutron diffraction measurement method for reinforced concrete [4,9,10]. In the neutron diffraction method, the rebar strain and stress are calculated by fitting the diffraction profiles of neutrons irradiated during a specific time period [4–10]. In general, the longer the measurement time, the clearer the diffraction profiles of the neutrons measured in the experiment. The clearer the diffraction profile, the more accurately the diffraction angle and lattice spacing can be calculated, resulting in more accurate measured values. However, the measurement time was determined empirically in most previous neutron diffraction studies on reinforced concrete specimens. For this reason, the accuracy and reliability of the rebar stress measured in previous studies are still to be determined. Moreover, the accurate measurement of the rebar stress is important for understanding the bonding behavior. Therefore, it is necessary to clarify the accuracy of the measured rebar stress obtained using the neutron diffraction method to ensure the accuracy and reliability of the measurements using this method and to understand the bond behavior between the rebar and concrete in more detail. Furthermore, if the measurement time can be determined from the allowable accuracy of the rebar stress, it is important to ensure the accuracy and reliability of the measurement results.

In this study, the relationship between measurement time and accuracy was examined for reinforced concrete specimens with different cross-sectional shapes, measurement positions, and cover thicknesses to clarify the accuracy of the measurements using neutron diffraction methods. This study provides new insight into the measurement accuracy of the rebar stresses inside the concrete as measured by neutron diffraction methods, which enhances our understanding of the applicability of this method in the field of concrete engineering.

2. Materials and Methods

2.1. Concrete Materials Used and Mix Proportion

Table 1 lists the materials used for the concrete in this experiment, and Table 2 lists the concrete mix proportions. Table 2 shows the compressive strength, f_c, and static modulus of elasticity, E_c, on the test day (23 days after mixing and forming). The values on the test day were obtained under the same curing conditions as the specimens shown in Section 2.2.

Table 1. Concrete materials used in this study.

Materials	Types and Properties
Water (W)	Deionized water
Cement (C)	High early-strength Portland cement. Density: 3.14 g/cm^3
Fine aggregate (S)	Land sand from the Oi river. Absolute dry density: 2.58 g/cm^3
Coarse aggregate (G)	Crushed stone from Ome. Maximum particle size: 10 mm Absolute dry density: 2.66 g/cm^3
Chemical admixture (Ad)	Lignin sulfonate, oxycarboxylate, and polycarboxylic acid compounds

Table 2. Concrete mix proportions and properties.

W/C (%)	Unit Weight (kg/m^3)				Ad (g)	Slump * (cm)	Air (%)	f_c (Test Day *) (MPa)	E_c (Test Day *) (GPa)	f_c (28 Days *) (MPa)
	W	C	S	G						
60	175	294	850	950	C × 1.7%	19.6	4.7	39.4	23.8	41.1

* "Slump" is a parameter of the consistency of the concrete and was measured in accordance with JIS (Japanese Industrial Standards) A 1101 "Method of test for slump of concrete". * "Test day" is the day of the experiment of the neutron diffraction method. * "28 days" means that it is cured for 28 days under standard curing conditions.

2.2. Experimental Parameters and Test Specimens

Figure 1 presents an overview of the specimens used in the experiment. The experimental parameters are presented in Table 3. Because the neutron permeability of concrete is lower than that of aluminum, the cross-sectional shapes of the specimens and cover thickness were set as experimental variables to qualitatively evaluate the effects of concrete and aluminum in the neutron transmission path on the measurement accuracy of the rebar stress. In addition, in specimen Nos. 1 and 3, the moisture loss from the concrete surface was suppressed by the aluminum pipe during the drying process, which will be described later. Therefore, in specimen Nos. 1 and 3, the moisture content in the concrete was non-uniform because of the drying from the two ends, the loaded and free ends (as shown in Figure 1). The measurement position was set as an experimental variable to qualitatively evaluate the effect of the concrete moisture state on the rebar stress measurement accuracy. The terms "No." in Table 3 correspond to those shown in Figure 1.

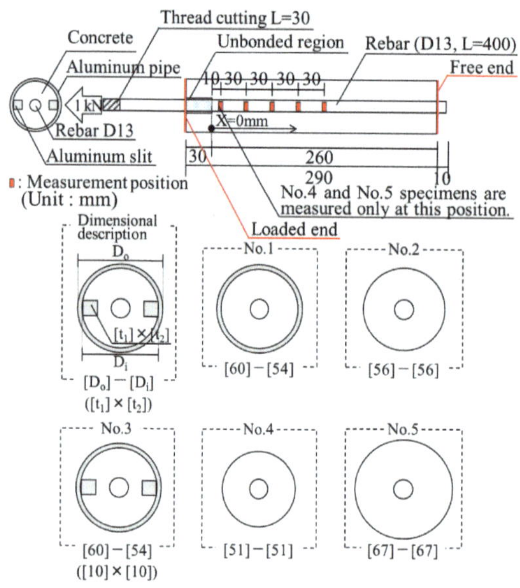

Figure 1. Schematic diagram of the specimens and measurement positions.

Table 3. Experimental parameters.

Series	Parameters
Series 1: Cross-sectional shape of the neutron path	No. 1, No. 2, No. 3 (as shown in Figure 1.)
Series 2: Measurement position (mm) (No. 1, No. 2, No. 3)	10, 40, 70, 100, 130 (as shown in Figure 1.)
Series 3: Cover thickness (mm)	19.0 (No. 4), 21.5 (No. 2), 27.0 (No. 5)

The specimens were prepared by placing a 400 mm-long rebar in concrete with the cross-sectional shape shown in Figure 1. The rebar was bonded with concrete over a length of 260 mm, and a 30 mm unbonded region was provided on the loading end. The unbonded region was created by removing the installed rubber hose at the time of casting after curing. The rebar used in this experiment was a commercially available D13 deformed bar (SD295, carbon steel) based on JIS G 3112 "Steel bars for concrete reinforcement".

For the specimens unconstrained by an aluminum sleeve (specimen Nos. 2, 4, and 5), the concrete specimen was cast in a mold made of PVC pipe and demolded after hardening. The specimens constrained by an aluminum sleeve (specimen Nos. 1 and 3) were prepared by using the aluminum pipe as the mold. The aluminum slit in specimen No. 3 was installed by attaching a 10 mm × 10 mm aluminum square pillar to the aluminum pipe with a room-temperature curing adhesive before concrete casting. When aluminum and concrete come into contact, there is concern that hydrogen bubbles may form in the concrete [11]. Therefore, a two-component modified epoxy resin paint was sprayed on the contact surface of the aluminum pipe to suppress chemical reactions.

The specimens were demolded 48 h after concrete casting and cured in water at 20 °C for 7 days. The specimens were then dried at 20 °C and 60% RH for 24 h. The specimens were dried in the constant temperature chamber at 40 °C for 11 days of aging and at 60 °C for a further 20 days of aging. After drying, the specimens were sealed with aluminum tape to prevent moisture absorption. A neutron diffraction method was used over four days, beginning on aging day 21. One specimen was used for each parameter in the neutron diffraction method.

2.3. Experimental Method

2.3.1. Overview of Measuring Rebar Strain

An angular dispersion type of neutron diffraction experiment was performed using the diffractometer for residual stress analysis (RESA) at the Japan Research Reactor No. 3 (JRR-3) of the Japan Atomic Energy Agency (JAEA). Neutrons are scattered when they hit individual atoms, and diffraction occurs when scattered neutrons interfere with each other if the Bragg diffraction condition in Equation (1) is satisfied:

$$2d \sin\theta = n\lambda \tag{1}$$

where d is the lattice spacing, θ is the diffraction angle, n is the diffraction order, and λ is the wavelength of incident neutrons.

The lattice spacing of the crystal lattice changes when a load is applied to the material. The change in the lattice spacing can be translated into a change in the diffraction angle by differentiating Equation (1) to form Equation (2).

$$\varepsilon = \frac{d - d_0}{d_0} = \frac{2\theta - 2\theta_0}{2} \cdot \cot\frac{2\theta_0}{2} \tag{2}$$

where ε is the elastic strain, d_0 is the lattice spacing in the initial state, and θ_0 is the diffraction angle in the initial state.

The elastic strain can be calculated by measuring the change in diffraction angle $\Delta\theta$.

An overview of the RESA is shown in Figure 2. Thermal neutrons of a single wavelength were extracted from the research reactor, and neutrons shaped by an incident Cd slit irradiated the sample. The neutrons diffracted by the sample were detected using a one-dimensional ^{3}He detector. The detector measured the diffraction intensity of each diffraction angle of neutrons for a certain period (5 min in this experiment). Assuming that the relationship between the diffraction angle and the diffraction intensity follows a Gaussian distribution and that the peak diffraction angle 2θ is obtained by fitting. From this peak diffraction angle, the lattice spacing d is obtained using Equation (1), and the elastic strain, ε, is calculated using Equation (2).

The actual area to be measured by the RESA is the square column area (gauge volume) shown in Figure 2, which is determined by the size of the incident Cd slit and the width of the radial collimator. The elastic strain was calculated using the average value of the volume. The diffraction plane used in this study was the (110) plane. The wavelength of the incident neutrons was 1.72 Å, the size of the incident Cd slit was 5 × 10 mm, and the width of the radial collimator was 5 mm.

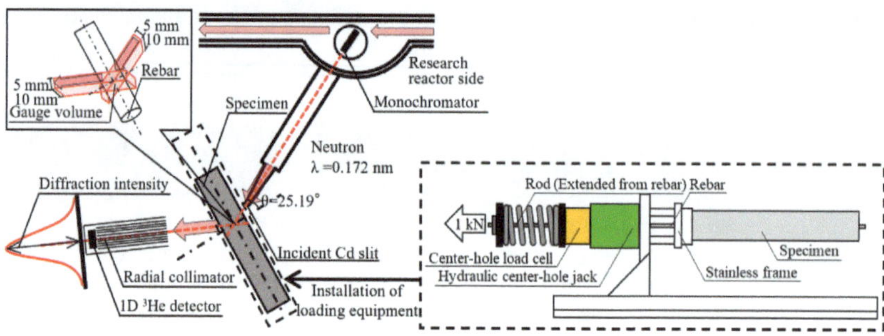

Figure 2. Schematic diagram of strain measurement of rebar by the RESA and loading equipment used in this experiment.

2.3.2. Loading Method and Measurement Position

Figure 2 presents an overview of the loading machine used in this experiment. A load cell and disc spring were placed on a hydraulic jack fixed to the loading machine, and one side of the specimen was fixed to the load machine through a rod extended from the rebar. A tensile force was introduced into the rebar by pushing the spring through the jack.

Figure 1 shows the measurement positions of the rebar stress for specimen Nos. 1–3. Rebar stress measurements were performed at five positions: 10, 40, 70, 100, and 130 mm from the beginning of the bonded region (0 mm). The measurement position of 130 mm in specimen No. 1 was not measured. For specimen Nos. 4 and 5, measurements were taken only at one measurement position, 10 mm from the beginning of the bonded region. The measurement results in Sections 3.1 and 3.3 were all taken at the 10 mm measurement position.

Table 4 lists the total measurement times for each measurement position. The total measurement time per measurement position was set for each specimen so that the lattice spacing (d_a), which is assumed to be the actual value, is sufficiently accurate, as described in Section 2.4.1. The applied load was set to 1 kN for all specimens to avoid any misalignment of the specimens during the measurement and the generation of large stresses at the measurement position.

Table 4. Total measurement time.

Specimen No.	Measurement Time (min)
No. 1	210 (5 × 42 times) [10, 40, 70, 100 mm]
No. 2	150 (5 × 30 times) [10, 40, 70, 100, 130 mm]
No. 3	120 (5 × 24 times) [10, 40, 70, 100, 130 mm]
No. 4	120 (5 × 24 times) [10 mm]
No. 5	180 (5 × 36 times) [10 mm]

The value in [ ] indicates the measurement position.

2.4. *Analysis Methods*

2.4.1. Variation of Rebar Stress Calculation

The analytical methods used in this study are described below. In this experiment, n measurements of 5 min each were made up to the times shown in Table 4 at each measurement position. Figure 3 shows an example of the relationship between the diffraction angle and diffraction intensity obtained from a 5-min measurement. The upper part of the figure shows an approximate Gaussian curve fitting equation analyzed using the graphing software Igor Pro. In this experiment, the peak diffraction angle (2θ), peak intensity (PI), and background intensity (BG) were obtained by fitting the relation between the diffraction angle and diffraction intensity obtained from a 5-min measurement (shown in Figure 3).

For measurement results longer than 10 min, shown in Sections 3 and 4 below, 2θ was calculated by averaging the 5-min measurement result over n times. The *PI* and *BG* values were calculated by adding the results over n times.

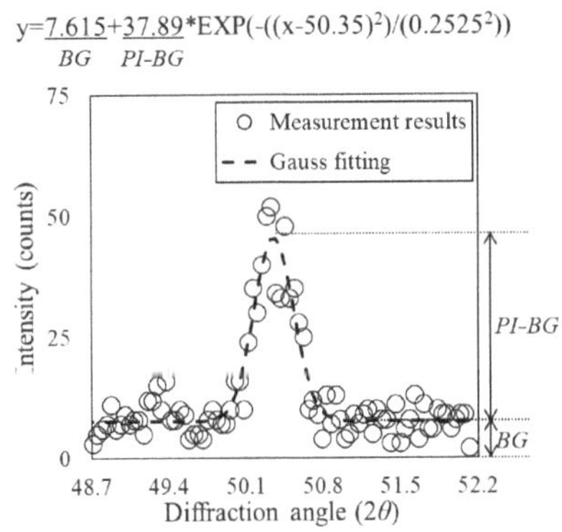

Figure 3. Relation between diffraction angle and intensity.

This experiment aimed to clarify the accuracy of rebar stress measurements using the neutron diffraction method. Therefore, in this experiment, the variation in stress at a certain measurement time was examined under the assumption that the lattice spacing obtained at the total measurement time shown in Table 4 is d_0, as shown in Equation (2). Assuming that the lattice spacing obtained at the total measurement time is the actual value (d_a), and substituting the lattice spacing obtained at a certain measurement time (Δt) for $d_{\Delta t}$ and the lattice spacing obtained at the total measurement time for d_a, the variation in strain at a certain measurement time (ε_{err}) is calculated using Equation (3). Figure 4 shows an image of the lattice strain variation during this experiment. In this experiment, the variation of rebar stresses was calculated by multiplying the strain variation by the diffraction elastic constants of the rebar (20.0 GPa). Although to determine the diffraction elastic constants, it is necessary to measure the strain under the application of the known uniaxial stress [12], in this experiment, the stress was calculated assuming the diffraction elastic constants of 20.0 GPa.

$$\varepsilon_{err} = \frac{d_{\Delta t} - d_a}{d_a} \tag{3}$$

where ε_{err} is the strain variation, $d_{\Delta t}$ is the lattice spacing obtained at a certain measurement time, and d_a is the lattice spacing obtained during the total measurement time.

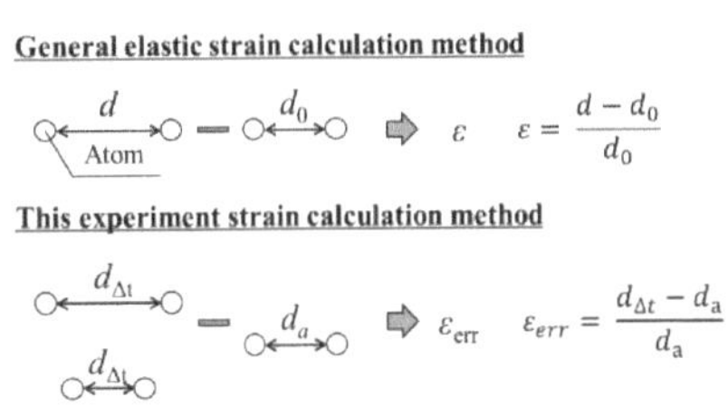

Figure 4. Image of the lattice strain variation in this experiment.

Figure 5 shows, as an example, the relation between measurement time and rebar stress obtained at the 10 mm position of specimen No. 3. The rebar stress shown in the figure was calculated by multiplying the diffraction elastic constants of the rebar by ε_{err} calculated from Equation (3). Sections 3 and 4 below discuss the experimental results by investigating the standard deviation (*SD*) of the variation of the rebar stress shown in Figure 5.

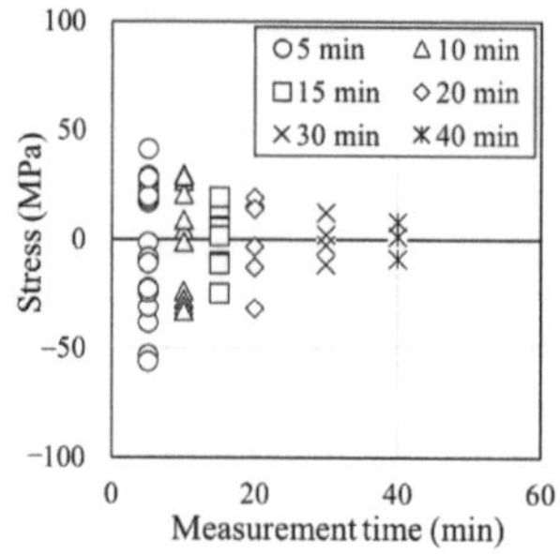

Figure 5. Relation between measurement time and stress ($\sigma = E \cdot \varepsilon_{err}$), (No. 3, 10 mm).

2.4.2. Analysis Method Sensitivity

Section 2.4.1 outlines the method for calculating the *SD* of the measured rebar stress, which is discussed in Section 3 below. However, in neutron diffraction method measurements, data for 30 min are rarely obtained in the form of six measurements of 5 min each. In general, the relation between diffraction angle and diffraction intensity is more often obtained as one measurement over 30 min [4,9,10]. Therefore, in this section, the relationship between measurement time and *SD* was calculated for the 10 mm measurement positions of specimen Nos. 2 and 3 according to the analysis method described above (Ave). The relation between the diffraction angle and diffraction intensity obtained from the 5-min measurement was then added *n* times to create a single dataset. The relation between measurement time and *SD* was then calculated (Sum). The influence of the analysis method on the measurement results was examined by comparing Ave and Sum.

Figure 6 shows the relationship between the measurement time and the *SD*. The figure shows that the measurement results were similar for the two analysis methods. Therefore, in Section 3, the measurement accuracy was examined further using the analysis method described above in Section 2.4.1 (Ave).

Figure 6. Relation between measurement time and *SD*: (**a**) specimen No. 1, 10 mm; and (**b**) specimen No. 3, 10 mm. *SD* stands for standard deviation and was calculated according to Section 2.4.1.

3. Measurement Accuracy Results

3.1. Cross-Sectional Shape Sensitivity

Figure 7 shows the relationship between the measurement time and the diffraction intensity for different cross-sectional shapes, and Figure 8 shows the relationship between the measurement time and the *SD*. Figures 9 and 10 show the same relationships, respectively, for different measurement positions. Figures 11 and 12 show the same relationships, respectively, for different cover thicknesses. Figures 8, 10 and 12 also show the power approximation equation.

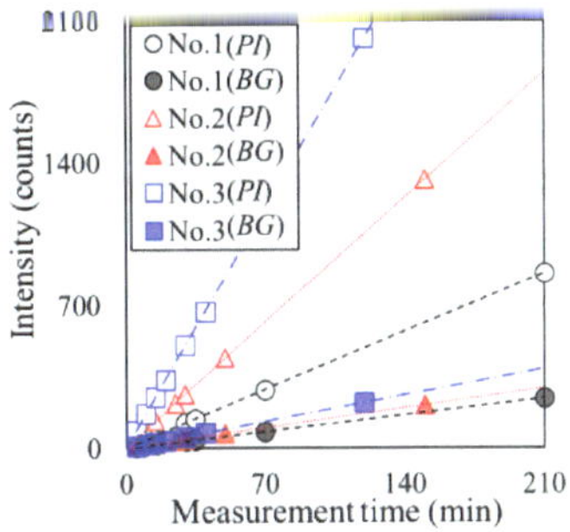

Figure 7. Measurement time and intensity in Series 1 (cross-sectional shape).

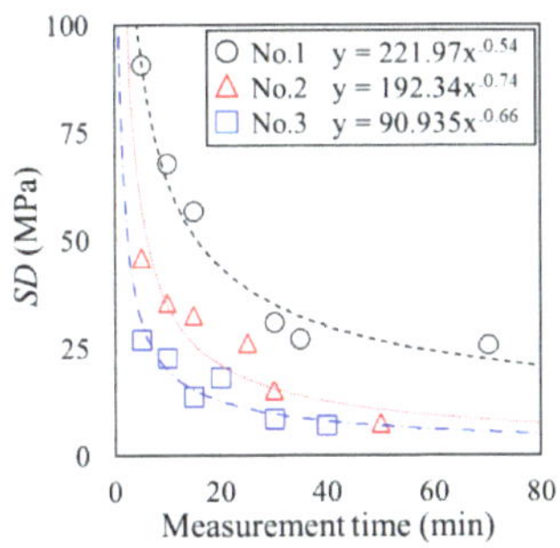

Figure 8. Measurement time and *SD* in Series 1 (cross-sectional shape).

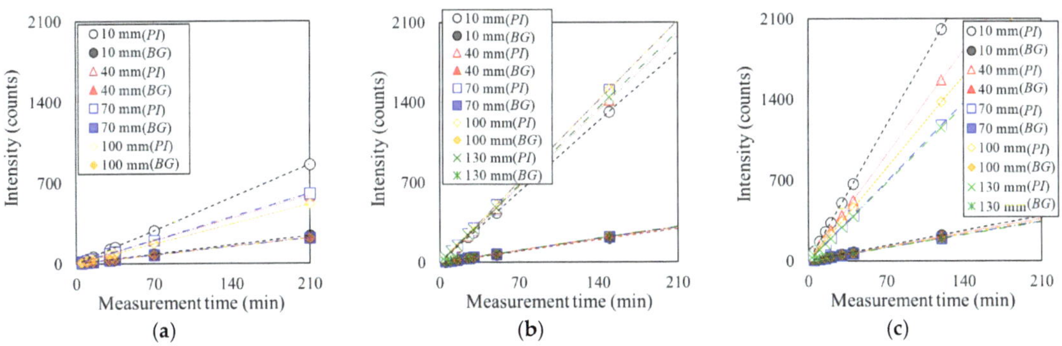

Figure 9. Measurement time and intensity in Series 2 (measurement position): specimens (**a**) No. 1, (**b**) No. 2, and (**c**) No. 3.

As shown in Figure 7, the rate of increase in diffraction intensity with increasing measurement time increases in the specimen order of No. 3, No. 2, and No. 1. Figure 8 also shows that specimen No. 3 has better measurement accuracy than the other specimens, even over a relatively short measurement time.

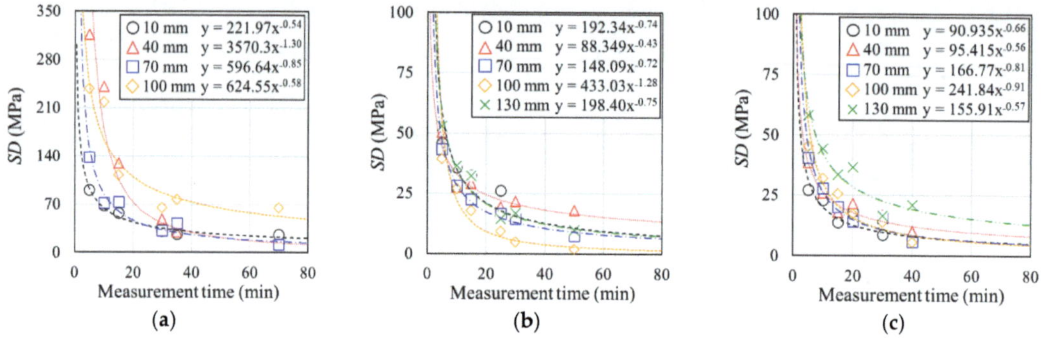

Figure 10. Measurement time and *SD* in Series 2 (measurement position): specimens (**a**) No. 1, (**b**) No. 2, and (**c**) No. 3.

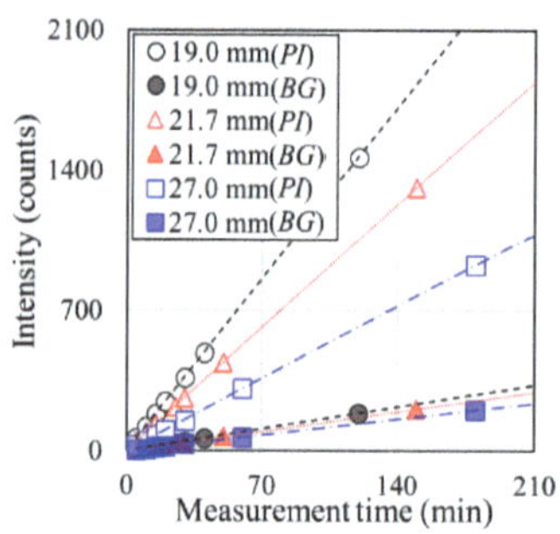

Figure 11. Measurement time and intensity in Series 3 (cover thickness).

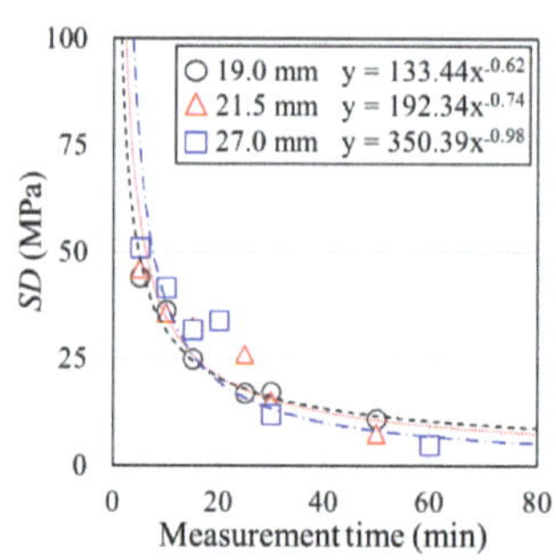

Figure 12. Measurement time and *SD* in Series 3 (cover thickness).

3.2. Sensitivity to Measurement Position

Focusing on the results of specimen Nos. 1 and 3, the rate of increase in intensity with increasing measurement time tends to decrease as the *PI* decreases from 10 mm to 130 mm (shown in Figure 9). Figure 10 also shows that the closer the measurement position is to 10 mm, the higher the measurement accuracy that can be obtained in a shorter measurement time. However, focusing on the value of the vertical axis in Figure 10, it can be seen that specimen No. 1, where the rate of increase in intensity is small, has a significant variation in rebar stress, even at the same measurement time as specimen Nos. 2 and 3.

Focusing on the result of specimen No. 2, the effect of the measurement position on the diffraction intensity is small. Furthermore, the difference in measurement accuracy due to the measurement position was smaller than that of specimen Nos. 1 and 3.

3.3. Sensitivity to Cover Thickness

As shown in Figure 11, the increase in *PI* and *BG* with increasing measurement time was more significant for specimens with a smaller cover thickness. Figure 12 also shows that the effect of the cover thickness on the measurement accuracy is small compared with the other experimental parameters.

4. Discussion

4.1. Effects of Different Factors on Diffraction Intensity

In reinforced concrete specimens, neutron attenuation occurs primarily because of the hardened cement in the concrete (CSH, Ca $(OH)_2$, etc.) and water (H_2O) in the voids. This is because the mass attenuation coefficient of neutrons is substantial for H atoms [9,13,14]. Therefore, in this experiment, the increase in diffraction intensity with measurement time was much more significant for specimen Nos. 3 and 4, where the amount of hardened cement in the neutron transmission path is small (as shown in Figures 7 and 11).

For specimens constrained by aluminum sleeves, such as specimen Nos. 1 and 3, the drying of the concrete proceeds only from two sides: the loaded and free ends. This is because the aluminum pipe intercepts water loss during the drying process. Therefore, in specimen Nos. 1 and 3, the moisture content gradient is due to drying on the two sides. For this reason, differences in the diffraction intensities due to the measurement position were observed for specimen Nos. 1 and 3 (as shown in Figure 9).

4.2. Effect of Diffraction Intensity on Accuracy

The variations in the rebar stress with respect to the measurement time tend to be smaller at the measurement positions where the increase in intensity is significant, such as specimen No. 3 at the 10 mm measurement position in Figure 9c. In addition, as shown in Figure 9a, the variation in rebar stress with measurement time tends to be more significant for specimens with a smaller increase in intensity than for the other specimens. Therefore, the *SD* of the rebar stress was affected by its intensity at that measurement time. However, if the increase in *BG* is as significant as the increase in *PI*, the Gaussian distribution shown in Figure 3 is also expected to be unclear. Therefore, in this section, the relationship between the intensity of the difference between *PI* and *BG* ($PI-BG$) and *SD* is discussed.

Figure 13 shows the relationship between $PI-BG$ and *SD* for all the measurement points obtained in Section 3. The figure shows the approximation equation and the coefficient of determination for the power approximation. Figure 14 shows the same information with the results from a previous study [15] and without a concrete cover added (i.e., for the rebar-only). Figure 13 shows that the *SD* tends to decrease as $PI-BG$ increases, suggesting that the relationship between the two is unique. On the other hand, the *SD* in this analysis is calculated by assuming the result of the total measurement time to be the actual value (see Section 2.4). At a measurement position where the $PI-BG$ at the total measurement time is insufficient, there is concern that the reliability of the d_a measurement value itself, which is assumed to be the actual value, may decrease. Therefore, we examined the effect of the total $PI-BG$ on the power-approximation coefficient of determination.

Figure 15 shows the relationship between the range of $PI-BG$ at the total measurement time and the coefficient of determination. The figure shows that the coefficient of determination of the power approximation is stable when only measurement positions with $PI-BG > 600$ counts at the total measurement time are analyzed. Kanematsu et al. [10] have previously reported that the measured stress shows a stable trend when $PI > 700$ counts, and a $PI > 700$ counts is equivalent to $PI-BG > 600$ counts in this analysis. Therefore, it can be inferred that the *SD* at the measurement positions where $PI-BG$ at the total measurement time >600 counts is a reliable value in this analysis.

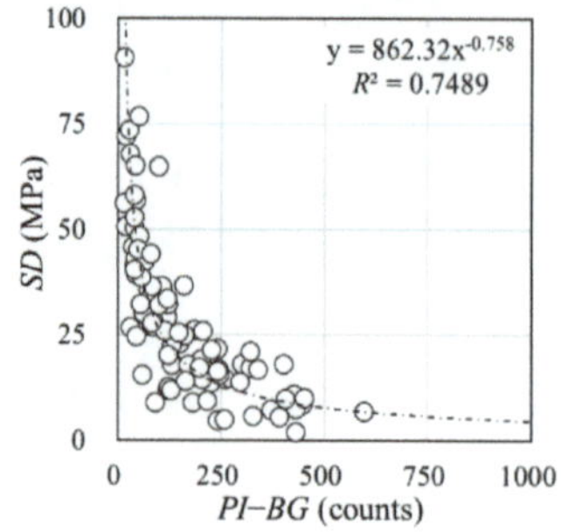

Figure 13. Relation between *PI−BG* and *SD* (All measurement positions).

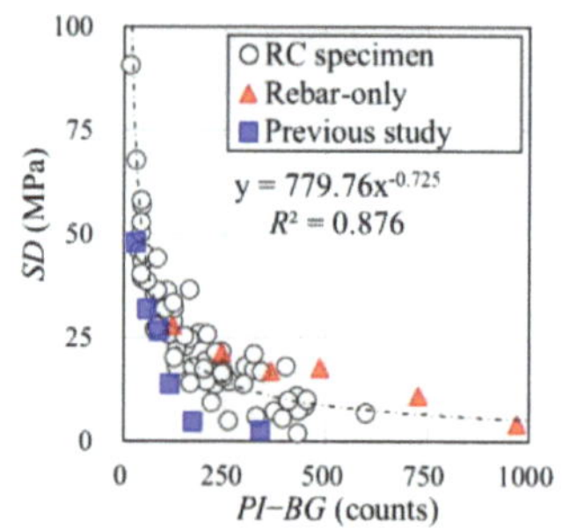

Figure 14. Relation between *SD* and *PI−BG* (total *PI−BG* $\geq$ 600 counts).

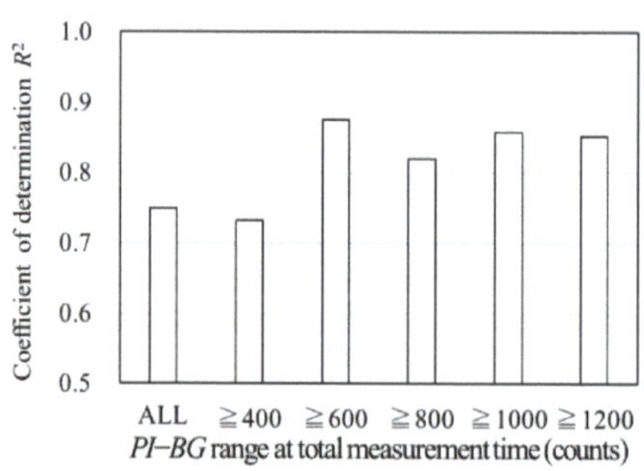

Figure 15. Relation between R^2 and the *PI−BG* range.

Figure 14 shows the relationship between *PI−BG* and *SD* at the measurement position where the total *PI−BG* > 600 counts. In addition, in a previous study [15], a specimen with the same cross-sectional shape as specimen No. 2 in this experiment was irradiated for a series of 36 measurements of 5 min each. Although the diffractometer, diffraction plane, slit size, and radial collimator are the same as in this experiment, the neutron wavelength in the previous study was 2.08 Å, which is >1.72 Å of this study. Figure 14 also shows the relationship between *PI−BG* and *SD* obtained in the previous study [15]. In addition, the results of the measurement of the position without a concrete cover (rebar-only) in this experiment are also shown in the figure. The rebar-only measurements were performed 24 times for 1 min. The figure shows that the relationship between *PI−BG* and *SD* is unique and that the coefficient of determination of the power approximation equation is high. In addition, the relationship between *PI−BG* and *SD* for rebar-only, where the diffraction intensity and total *PI−BG* are very high, shows a similar trend to the power approximation equation in Figure 14. This indicates that, under the same optical setup as in this experiment, the power approximation equation shown in Figure 14 is valid for reinforced concrete specimens under different conditions.

The relationship between $PI-BG$ and SD in the previous study [15] had a higher measurement accuracy than in this experiment at the same $PI-BG$. It is inferred that the wavelength of the neutrons used in this experiment, λ, influences this. Focusing on Equation (1), it can be observed that as λ increases, the change in θ with respect to the change in d increases. Therefore, it can be inferred that in the previous study [15], λ is more significant than that in this experiment, and the measurement accuracy is better. This is because the change in θ was measured at a higher resolution. However, as shown in Equation (4), the slope of $PI-BG$ becomes more extensive because the energy is more significant when λ is small.

$$\lambda = \frac{0.9045}{\sqrt{U}} \tag{4}$$

where U is the energy (meV).

4.3. Measurement Accuracy Verification

Section 4.2 presented a unique relationship applicable under the same optical setup as that in this experiment. However, the equation in Figure 14 is expected to change when the optical setup, such as neutron wavelength, diffraction plane, slit size, and radial collimator width, is different. This equation is also expected to change when the target materials are different. Therefore, when applying the neutron diffraction method for rebar stress measurement, it is necessary to calculate the relationship between $PI-BG$ and SD for each optical setup and the material to be measured and to clarify the variation in rebar stress caused by the experiment.

Figure 16 shows an example of determining the measurement time for the rebar stress measurements. In this experiment, the relationship between the $PI-BG$ and SD was calculated mainly for positions with a concrete cover. However, as described in Section 4.2 above, similar measurement results can be obtained for the rebar-only measurement (as shown in Figure 14). The measurement time for the rebar-only measurement can be significantly reduced. This is because the diffraction intensity is exceptionally high compared with the concrete cover position. Therefore, in the procedure proposed in Figure 16, the accuracy was verified from the measurement results obtained for the rebar alone.

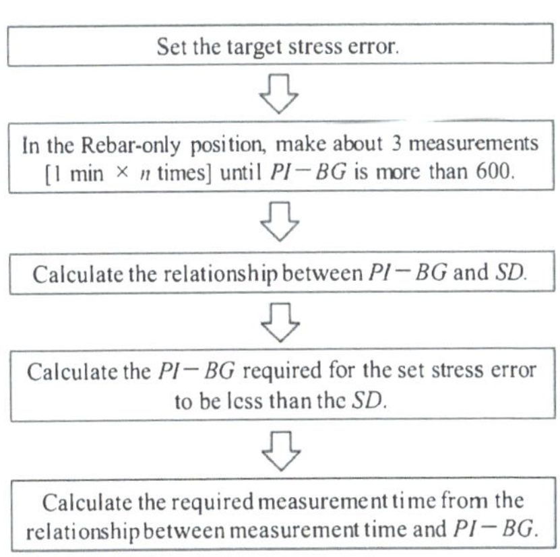

Figure 16. Method of determining measurement time (as shown in Figures 7, 9, 11, and 14).

Following the procedure shown in the figure, the required measurement time was calculated from the allowable SD to ensure the accuracy and reliability of the measured rebar stress. Note that the value of $PI-BG$ of 600 counts or more, as shown in Figure 16, is based on the condition of this experiment; therefore, it is necessary to set an appropriate value on the safe side for applications.

5. Conclusions

This study examined the measurement accuracy of rebar stress obtained by the neutron diffraction method using reinforced concrete specimens under different conditions of continuous neutron irradiation for 5 min. The following findings were obtained from the neutron diffraction intensities and *SD*s of the rebar stress in this study:

1. The increase in diffraction intensity with increasing measurement time is more significant because of the installation of aluminum slits and decreasing cover thickness, resulting in decreased hardened cement and water contents in the neutron transmission path.
2. For specimens where the increase in diffraction intensity with an increase in measurement time is significant, the measurement accuracy tends to be high in short-time measurements.
3. Under the conditions of this experiment, the analytical results tend to be stable when $PI-BG > 600$ counts.
4. The *SD* of the rebar stress decreases as $PI-BG$ increases, and the relationship between the two values is expressed by a power approximation equation.
5. By calculating the required measurement time from the *SD* of the rebar stress following the procedure in Figure 16, it is possible to ensure the reproducibility and reliability of the rebar stress even for reinforced concrete specimens with different measurement conditions.

Author Contributions: Conceptualization, A.Y. and M.K. (Manabu Kanematsu); Investigation, A.Y. and S.M.; Methodology, A.Y., K.K. and M.K. (Manabu Kanematsu); Formal analysis, A.Y.; Project administration, A.Y. and M.K. (Manabu Kanematsu); Resources, A.Y., M.K. (Mayu Kawakami) and S.M.; Validation, A.Y. and S.M.; Supervision, M.K. (Manabu Kanematsu); Visualization, A.Y.; Writing—original draft, A.Y., J.K., Y.N. and M.K. (Manabu Kanematsu); Funding acquisition, Y.M., Y.N., T.M., S.M. and M.K. (Manabu Kanematsu). All authors have read and agreed to the published version of the manuscript.

Funding: This research was funded by JSPS KAKENHI, grant number JP20H02302.

Data Availability Statement: No new data were created or analyzed in this study. Data sharing is not applicable to this article.

Acknowledgments: This work was performed under the Shared Use Program of JAEA Facilities (Proposal No. I537).

Conflicts of Interest: The authors declare no conflict of interest.

References

1. Nilson, A.H. Internal Measurement of Bond Slip. *ACI J. Proc.* **1972**, *69*, 439–441. [CrossRef]
2. McHenry, D.E.; Walker, W.T. Laboratory Measurements of Stress Distribution in Reinforcing Steel. *ACI J. Proc.* **1948**, *44*, 1041–1054. [CrossRef]
3. Shima, H.; Chou, L.L.; Okamura, H. Micro and Macro Models for Bond in Reinforced Concrete. *J. Fac. Eng. Univ. Tokyo* **1987**, *39*, 133–194.
4. Suzuki, H.; Kanematsu, M.; Kusunoki, K. Neutron Diffraction Studies on Strain Evaluation of Rebar in Reinforced Concrete. *Powder Diffr.* **2009**, *24*, S68–S71. [CrossRef]
5. Suzuki, H.; Kusunoki, K.; Hatanaka, Y.; Mukai, T.; Tasai, A.; Kanematsu, M.; Kabayama, K.; Harjo, S. Measuring Strain and Stress Distributions Along Rebar Embedded in Concrete Using Time-of-Flight Neutron Diffraction. *Meas. Sci. Technol.* **2014**, *25*, 025602. [CrossRef]
6. Suzuki, H.; Kusunoki, K.; Kanematsu, M.; Tasai, A.; Hatanaka, Y.; Tsuchiya, N.; Bea, S.; Shiroishi, S.; Sakurai, S.; Kawasaki, T.; et al. Application of Neutron Stress Measurement to Reinforced Concrete Structure. In Proceedings of the 2nd International Symposium on Science at J-PARC, Ibaraki, Japan, 12–15 July 2014. [CrossRef]

7. Suzuki, H.; Kusunoki, K.; Kanematsu, M.; Mukai, T.; Harjo, S. Structural Engineering Studies on Reinforced Concrete Structure Using Neutron Diffraction. In Proceedings of the International Conference on Residual Stresses (ICRS) 2016, Sydney, Australia, 3–7 July 2016. [CrossRef]
8. Kobayashi, K.; Suzuki, H.; Nishio, Y.; Kanematsu, M. Evaluation of Bond Performance of Reinforced Concrete Using Hot-Dip Galvanized Rebar by Neutron Diffraction. *J. Struct. Constr. Eng. AIJ* **2021**, *86*, 1026–1035. (In Japanese) [CrossRef]
9. Kusunoki, K.; Kabayama, K.; Mukai, T.; Hatanaka, Y. Experimental Study on Bonding Action with Neutron Measuring System. *Proc. JCI* **2009**, *31*, 619–624. Available online: https://data.jci-net.or.jp/data_pdf/31/031-01-2104.pdf (accessed on 8 February 2023). (In Japanese)
10. Kanematsu, M.; Ota, T.; Suzuki, Y.; Noguchi, T. Neutron Diffraction Studies on Strain Evaluation of Rebar Around Cracks. *Proc. JCI* **2011**, *33*, 1853–1858. Available online: https://data.jci-net.or.jp/data_pdf/33/033-01-1303.pdf (accessed on 8 February 2023). (In Japanese)
11. Lin, Y.; Leong, B.S.; Hu, Z.-T.; Yang, E.-H. Autoclaved Aerated Concrete Incorporating Waste Aluminum Dust as Foaming Agent. *Constr. Build. Mater.* **2017**, *148*, 140–147. [CrossRef]
12. Tanaka, K.; Akiniwa, Y.; Hayashi, M. Neutron Diffraction Measurements of Residual Stresses in Engineering Materials and Components. *Mater. Sci. Res. Int.* **2002**, *8*, 165–174. [CrossRef] [PubMed]
13. Kanematsu, M.; Maruyama, I.; Noguchi, T.; Tsuchiya, N. Quantification of Water Penetration into Concrete through Cracks by Neutron Radiography. *Nucl. Instrum. Methods Phys. Res. Sect. A Accel. Spectrometers Detect. Assoc. Equip.* **2009**, *605*, 154–158. [CrossRef]
14. Perfect, E.; Cheng, C.-L.; Kang, M.; Bilheux, H.Z.; Lamanna, J.M.; Gragg, M.J.; Wright, D.M. Neutron Imaging of Hydrogen-Rich Fluids in Geomaterials and Engineered Porous Media: A review. *Earth-Sci. Rev.* **2014**, *129*, 120–135. [CrossRef]
15. Kobayashi, K.; Yasue, A.; Kim, J.; Nishio, Y.; Miyazu, Y.; Mukai, T.; Kanematsu, M. The Effect on the Measurement Time of Neutron Diffraction Method on the Accuracy of Measuring Rebar Stress. In Proceedings of the Summaries of Technical Papers of Annual Meeting, AIJ, Materials and Construction, Hokkaido, Japan, 5–9 September 2022. (In Japanese)

quantum beam science

MDPI

Article

Measurement of Mechanical Behavior of ^{11}B-Enriched MgB_2 Wire Using a Pulsed Neutron Source

Shutaro Machiya [1,*], Kozo Osamura [2], Yoshimitsu Hishinuma [3], Hiroyasu Taniguchi [4], Stefanus Harjo [5] and Takuro Kawasaki [5]

1 Department of Mechanical Engineering, Daido University, 10-3 Takiharu-cho, Minami-Ku, Nagoya 457-8530, Japan
2 Resarch Institute for Applied Science, 49 Tanaka-ohi-cho, Sakyo-ku, Kyoto 606-8202, Japan
3 National Institute for Fusion Science, 322-6 Oroshi-cho, Toki-shi 509-5292, Japan
4 Osaka Alloying Works, Co., Ltd. 45-5-9 Shirakata-cho, Fukui-shi 910-3138, Japan
5 J-PARC Center, Japan Atomic Energy Agency, Tokai-Mura, Naka-Gun, Ibaraki 319-1195, Japan
* Correspondence: machiya@daido-it.ac.jp; Tel.: +81-52-612-6111 +2529

Abstract: MgB_2 represents a hexagonal superconductive material renowned for its straightforward composition, which has facilitated the development of cost-effective practical wires. Its capacity to function at temperatures as low as liquid hydrogen (LH_2) has made it a prominent candidate as wire material for the coils of next-generation fusion reactors. Much like other superconducting wires, a prevalent issue arises when these wires are employed in coils, wherein electromagnetic forces induce tensile stress and strain within the wire. This, in turn, diminishes the critical current, which is the maximum current capable of flowing within the generated magnetic field and strain. The techniques and methods for accurately measuring the actual strain on the filaments are of paramount importance. While strain measurements have been conducted with synchrotron radiation and neutrons for other practical wires in the past, no such measurements have been undertaken for MgB_2. Presumably, this lack of measurement is attributed to its relatively greater thickness, making it less suitable for synchrotron radiation measurements. Additionally, the high absorption cross-section of the included boron-10 poses challenges in obtaining elastic scattering data for neutron measurements. In response, we fabricated a wire enriched with boron-11, an isotope with a smaller neutron absorption cross-section. We then embarked on the endeavor to measure its strain under tensile loading using pulsed neutrons. Consequently, we succeeded in obtaining changes in the lattice constant under tensile loading through Rietveld analysis. This marks the inaugural instance of strain measurement on an MgB_2 filament, signifying a significant milestone in superconductivity research.

Keywords: superconductor; strain measurement; neutron scattering; MgB_2

1. Introduction

The MgB_2 superconducting compound was first discovered in 2001 [1]. MgB_2 possesses several notable characteristics, including a higher critical temperature (T_c) of 39 K, a simple binary chemical composition, lower specific gravity, and a relatively cost-effective production process. One of its key advantages is its usability at liquid hydrogen (LH_2) temperatures around 20 K. This makes it particularly well-suited for cable applications, thanks to its low strain and ease of use [2]. While most current coil applications fall within the low stress–strain range [3–5], the prospect of high-current applications like nuclear fusion necessitates an enhancement of the wire's strain properties, a critical concern. Notably, superconducting wire finds substantial use in fusion reactor projects. The ITER project, currently under construction, is slated to operate in a 4 K environment using copious amounts of liquid helium, referred to as LTS (low-temperature superconductor). However, the supply of liquid helium has experienced instability over several years, leading to a substantial increase in its price. This situation underscores the need for the next generation of fusion reactors to be helium-free for practicality reasons. Most contemporary fusion plans hinge on high-temperature superconductors (HTS) [6,7]. MgB_2 emerges as a promising candidate for helium-free fusion due to its compatibility with liquid hydrogen. However, when employed in coils for applications like nuclear fusion, the electromagnetic forces generated induce tensile stress or strain within the wire. Superconducting wires typically confront the issue of diminishing critical current, the maximum current-carrying capacity, under the influence of tensile or compressive strain. These effects, elucidated by Ekin as the Strain effect [8], originate from crystal deformation. In the case of Nb_3Sn, it is known that the strain-free state is most conducive to current flow. While Ekin was able to conduct strain-free experiments using large single crystals [8], practical wires necessitate quantum beam measurements to accurately gauge filament strain. Consequently, in composite superconducting wires, filament strain significantly influences performance under tension or compression, emphasizing the importance of non-destructive filament strain measurement. The measurement of real filament strain has been achieved in practical superconducting wires such as REBCO [9,10], Nb_3Sn [11], and BSCCO [12] through quantum beam experiments. However, this has yet to be reported for MgB_2 wires. In the design of high-field magnet coils, the prediction or measurement of actual filament strain for performance estimation constitutes a crucial element. It is anticipated that filament strain measurement technology will be indispensable for MgB_2 coil applications.

Natural boron comprises two isotopes: 20 wt% boron-10 (^{10}B) and 80 wt% boron-11 (^{11}B). Material stability holds paramount significance when considering its application in nuclear fusion [12]. Among these materials, ^{10}B has been identified as prone to decomposing into Li and He gases through nuclear reactions with neutrons. To circumvent this issue, research efforts are underway to develop superconductors with ^{11}B substitutions [13–15]. The ^{10}B isotope exhibits a substantial neutron absorption cross-section [16], rendering materials containing boron challenging to assess using neutron scattering. Consequently, no prior neutron scattering measurements have been conducted on MgB_2 wire. Although MgB_2 is intrinsically brittle, it possesses the capacity to withstand strains exceeding 0.2%. Given the numerous defects present in MgB_2 filaments, the reasons enabling practical use within this strain range remain largely unexplored. While oxide-type high-temperature superconductors experience current transport interruptions when crystal grain boundaries are disrupted, it has been suggested that MgB_2 superconducting wire may not be as affected by grain boundaries. Nevertheless, the extent to which the filament contributes to this aspect remains unknown. We believe that this can be conclusively ascertained through non-destructive measurements of the stress applied to the filament. To address this challenge, there is no alternative to experiments employing quantum beams. We conducted experiments at SPring-8 using high-energy X-rays. However, the large diameter of the MgB_2 wire posed difficulties in these measurements. Experiments using high-energy X-rays above 70 keV were infeasible, leaving neutron scattering as the sole viable option. Unfortunately, the original MgB_2 wire's neutron scattering experiment was infeasible due to the substantial neutron absorption cross-section of ^{10}B. Consequently, we opted to prepare ^{11}B-enriched MgB_2 wire to assess its filament's mechanical behavior. This ^{11}B-enriched wire was developed for fusion reactors by Hishinuma et al. Its primary purpose was to mitigate heat generation during neutron irradiation, a function dependent on its neutron absorption cross-section, while also preventing decomposition into lithium and helium through nuclear reactions. We also conducted neutron scattering experiments on commercially available practical MgB_2 wires to explore the possibility of MgB_2 diffraction. These measurements were carried out at Takumi (MLF BL19) of J-PARC [17], a neutron research facility at the Japan Atomic Energy Agency (JAEA). The measurement method employed at Takumi is the Time-of-Flight (ToF) method.

2. Materials and Methods

2.1. Preparation of MgB_2 Wires

Figure 1 displays a multi-filamentary MgB_2 wire enriched with ^{11}B. The wire has a diameter of approximately 1.07 mm. This ^{11}B-enriched sample was developed by the National Institute for Fusion Science (NIFS) in Japan. The cracks seen in the photographs are a result of lateral compression near the chuck of the specimen used for tensile testing and are not representative of the wire's typical condition. The manufacturing process is outlined below.

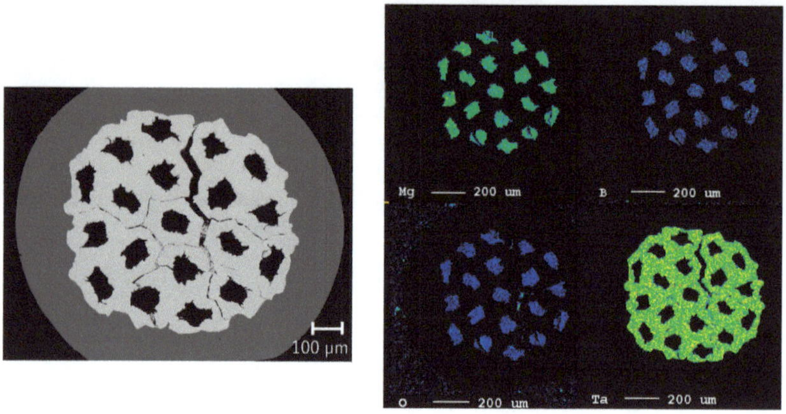

(a) (b)

Figure 1. Cross-section SEM image of NIFS ^{11}B-enriched MgB_2 wire: (**a**) secondary electron image; (**b**) color map of EPMA results (chemical symbols: Mg, B, O and Ta).

Mg powder (purity: 99.9%, −200 mesh) and ^{11}B isotopic powders, which were isotopically separated from natural boron, were prepared. The ^{11}B powder was sourced from Ceradyne and had a purity of 98%. The Mg_2Cu intermetallic compound was used as the source of Cu additive [18]. To ensure uniform dispersion within the precursor powder, microparticulation was employed. In this study, Mg_2Cu microparticles were produced through mechanical milling using a ball mill. The amount added was equivalent to 3 at% Cu, which was considered optimal for the precursor powder. The powder-in-tube (PIT) method was utilized to fill a metallic tantalum tube with an outer diameter of 10 mm and an inner diameter of 6 mm. In the subsequent step, 19 hexagonal single-core wires were cut and assembled into oxygen-free copper tubes (outer diameter: 14 mm, inner diameter: 10 mm). Finally, this composite billet was drawn using a cassette roller die, resulting in a final diameter of 1.07 mm to obtain the multi-core wire. In the last stage, the heat treatment temperature was raised from 450 to 600 degrees Celsius over a period of 200 h in an argon atmosphere to facilitate low-temperature diffusion. Figure 1a provides a cross-sectional view of the resulting wire via SEM, while Figure 1b displays the color map of EPMA results for the main elements. Some oxygen is observed in the EPMA results, which is typical in practical wires and is likely attributed to residual oxides.

Practical MgB_2 wires are commercially produced by several manufacturers using both ex situ and in situ methods. The filaments consist of sintered MgB_2 components and are enclosed by an inner sheath to prevent chemical reactions with the matrix. The outer sheath is crucial for ensuring consistent superconducting and mechanical properties over long-scale wires. In this particular case, we used wires produced by the Hypertech and Samdong in situ methods. These boron isotopes are assumed to be in their natural ratio. Table 1 illustrates the structural details of the two types of MgB_2 wires commonly used in practical applications. Additionally, Figures 2 and 3 display SEM cross-sectional images of these wires. The outermost sheath is a Ni-Cu-Fe alloy known as Monel, the darker region represents the MgB_2 filament, and each filament is surrounded by an Nb sheath. The diameter of each wire is 0.83 mm for both Samdong and Hypertech.

Table 1. Components of the commercial wires and their volume fraction.

Manufacturer	Filament	Filament 2	Inner Sheath	Matrix	Outer Sheath
Samdong	MgB_2 (0.132)		Nb (0.165)	Cu (0.320)	Ni-Cu-Fe alloy (0.382)
Hypertech	MgB_2 (0.099)	Cu (0.149)		Nb (0.369)	Ni-Cu-Fe alloy (0.382)

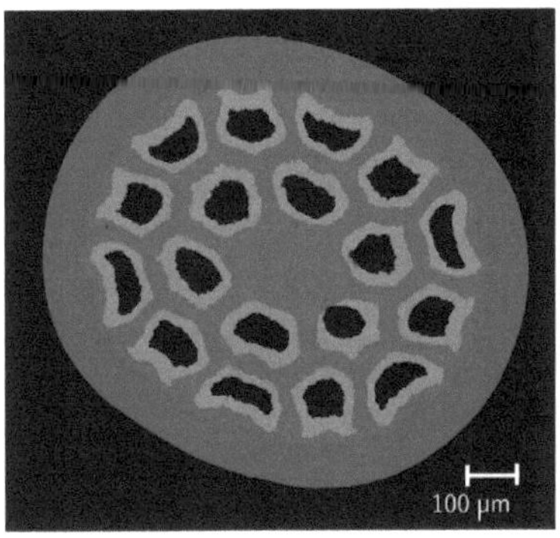

Figure 2. Cross-section SEM image of Samdong MgB_2 wire.

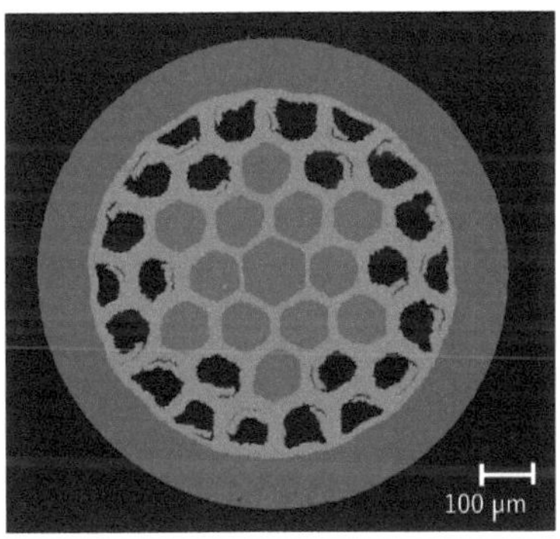

Figure 3. Cross-section SEM image of Hypertech MgB_2 wire.

2.2. Strain Calculation Using the ToF Method

The strains were derived from measurements conducted using the Time-of-Flight (ToF) method at Takumi (BL19 of MLF) [17]. The pulsed neutron method, also referred to as the ToF method, is a technique that captures the lattice spacing (d) of crystals in a histogram-like fashion, similar to a spectrometer. This method involves utilizing a single pulse of neutrons with varying speeds: faster neutrons diffract and are detected more rapidly, while slower neutrons are detected at a slower pace. In the case of Takumi, it can be conceptualized as a diffractometer with a 2θ angle fixed at approximately 90 degrees and -90 degrees. In practice, the precisely fixed 90-degree angle is less efficient for detection, so the angle is extended to encompass ±15 degrees. If 2θ is rigidly fixed to maintain a stringent scattering vector, fewer crystals will satisfy the diffraction conditions. Therefore, the ToF machine employs a method to enhance detection efficiency by widening the detector's range beyond 90 degrees. The relationship between lattice spacing (d) and flight time (t) in the Time-of-Flight method is elucidated by the following equation:

$$d = \frac{\lambda}{2\sin\theta} = \frac{1}{2\sin\theta} \cdot \frac{ht}{mL} \tag{1}$$

where (h) is Planck's constant and (m) is the mass of a neutron. (L) is a flight distance of neutrons from the chopper to the nearby target. From Equation (1), the lattice spacing (d) can be calculated using the flight times conversion parameter.

$$\varepsilon = \frac{d - d_0}{d_0} \tag{2}$$

Strain (ε) is determined through Equation (2). It is important to note that both d and ToF share the same dimensions, and this strain relationship remains valid whether we use d or ToF. In cases where there is no d_0 sample, which represents the strain-free d spacing (for instance, in tensile tests), the point at which zero load is applied is often calculated as d_0.

2.3. *In Situ Strain Measurement under Tensile Test and Equipment*

BL19 Takumi is a Time-of-Flight (ToF) neutron powder diffractometer designed for engineering science. What sets it apart from other powder scattering systems is its unique ability to precisely define the principal strain direction, aligning it with the scattering vector. As depicted in Figure 4, the detector comprises two opposing banks, North and South. When the sample is oriented at a 45° angle, the system can measure the lattice spacing (d) along the sample axis using the North bank and the lattice spacing (d) along the sample diameter using the South bank. It is worth noting that while the diffraction angle 2θ is theoretically fixed at 90 degrees, in practice, it is measured within the range of 75 to 105 degrees. This extended range allows for the measurement of diffraction from crystals with orientations that deviate by up to ±7.5° from the nominal principal strain direction. However, due to the nature of the sine function, the impact of such deviations is negligible.

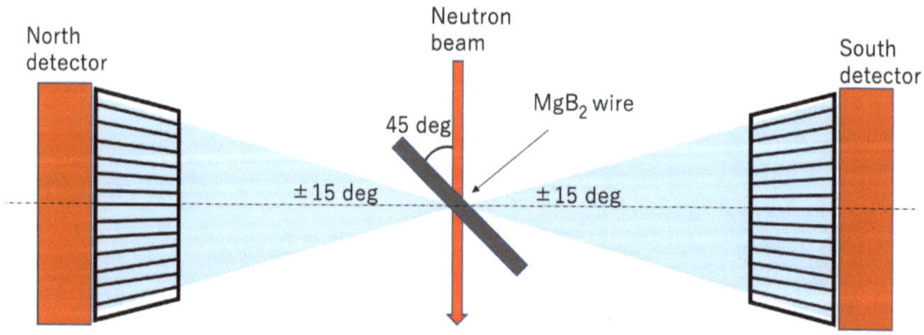

Figure 4. Schematic of the Takumi diffractometer (MLF BL19).

The detector's field of view is constrained by a radial collimator, which selectively captures scattering within a central 5 mm area while blocking out other contributions. An actual photograph of the diffractometer with the sample mounted is given in Figure 5. While diffraction appears to be primarily detected within the same beam plane, the actual detector is three-dimensional and can capture diffraction events within a solid angle ranging from −15 degrees to +15 degrees relative to the out-of-plane direction of the beam.

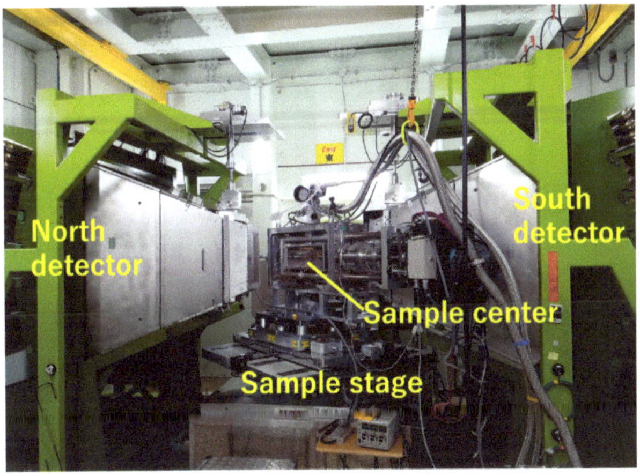

Figure 5. BL19 Takumi diffractometer of and sample.

The samples were affixed to a low-temperature tensile testing frame, although the measurements themselves were conducted solely at room temperature. The size of the neutron beam for irradiation was delimited by a slit, measuring 5 mm in length and 10 mm in width. The beam power generated by the MLF target was nominally set at 500 kW.

The wires used for the tensile tests were all 100 mm in length. Figure 6 displays a photograph of the actual sample setup. Although the tensile test frame was originally designed as a cryogenic tensile frame capable of conducting tests at temperatures around 10 K [19], these measurements were conducted at room temperature. The existing load cell was initially designed to accommodate large loads of up to 50 kN and therefore possessed excessive capacity, rendering it unsuitable for the current samples. Consequently, a smaller 2 kN load cell was custom-manufactured and positioned adjacent to the sample.

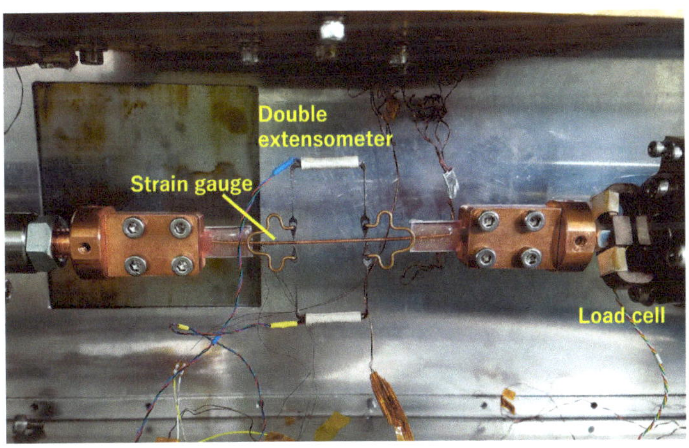

Figure 6. Sample with extensometer and load frame.

Engineering strain measurements were conducted using a Niylas-type double extensometer [20], known for its compactness, lightweight design, and precision. This type of extensometer is frequently employed for tensile measurements on superconducting wires and operates seamlessly at temperatures up to 4 K.

To secure the wire, epoxy adhesive was used to affix it to a 1 mm thick GFRP (glass fiber reinforced polymer) plate, which was subsequently clamped between copper chucks.

Additionally, to mitigate potential issues such as extensometer wire breakage, strain measurements were also carried out using custom-made strain gauges. These strain gauges were affixed to wires with a diameter of approximately 1 mm and were capable of measuring axial strain.

As illustrated in the image, even in the presence of neutrons, single-wire measurements were feasible. The ability to conduct experiments with a favorable signal-to-noise ratio (S/N) on a single wire owes much to the potent pulsed neutrons available at the J-PARC MLF. We previously conducted similar experiments using reactor-based steady-state neutron sources, but those experiments involved multiple wires [11]. In those cases, conducting experiments became challenging due to load-sharing variations among multiple wires. Considering that composite wires with a diameter of about 1 mm are employed to measure the strain of the ceramics contained within, it is reasonable to assume that conducting experiments with a single wire using a steady-state neutron source would be considerably more challenging.

Preliminary tensile tests were conducted before the experiments at the neutron facility. The stress–strain curve for the NIFS wire is depicted in Figure 7. In the stress–strain diagram, the horizontal axis represents the strain in percentage, calculated as the ratio of elongation, and it is dimensionless. The vertical axis represents the load in Newtons (N), measured using the load cell. Young's modulus can be determined from the initial slope of this curve, while the 0.2% proof stress can be found by offsetting a line with the same slope as Young's modulus to the right by 0.2% from the origin. To improve the accuracy of the Young's modulus measurement in accordance with IEC standards (IEC 61788-19:2013) [21] for superconducting wire, the load was briefly reduced to 0.15% strain. Three tensile tests were conducted, resulting in an average Young's modulus of 84 GPa and a 0.2% proof stress of 213 MPa. In this figure, the solid line represents a straight line with the same slope as Young's modulus, starting at 0.2% strain, and the point where this line intersects the stress–strain curve is the 0.2% proof stress.

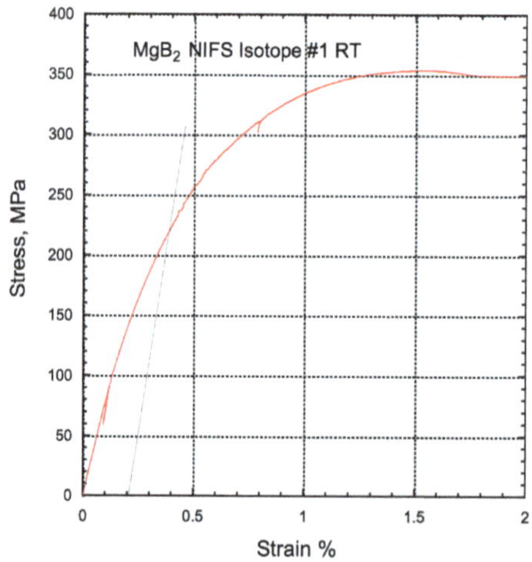

Figure 7. Stress–strain diagram of NIFS ^{11}B-enriched MgB_2 wire.

Hypertech and Samdong conducted similar tests, although the data for these tests are not presented here due to the lack of comparative neutron measurement data. Stress–strain curves for Hypertech and Samdong are shown in Figures 8 and 9, respectively. The average Young's modulus and 0.2% proof stress obtained for Hypertech were 133 GPa and 233 MPa, while for Samdong, they were 123 GPa and 734 MPa. The straight line in the diagram is the slope of Young's modulus and, as in Figure 7, is an auxiliary line for determining the 0.2% proof stress.

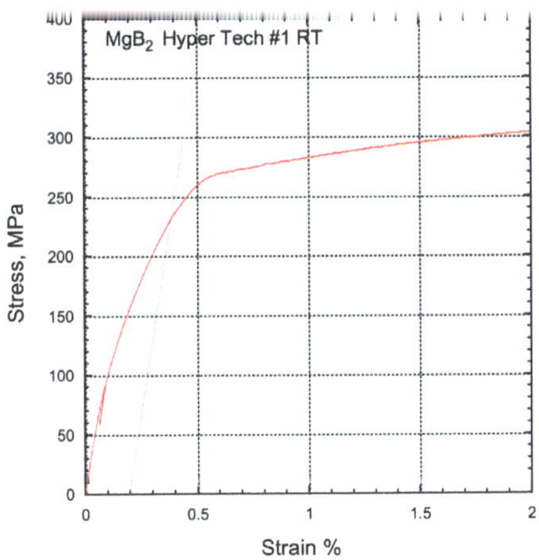

Figure 8. Stress–strain diagram of Hypertech MgB_2 wire.

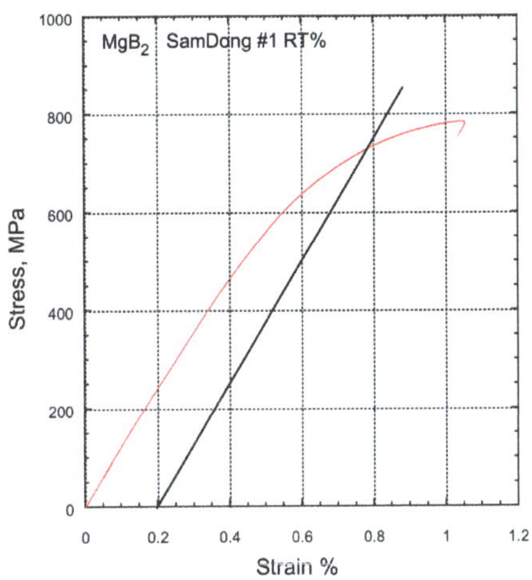

Figure 9. Stress–strain diagram of Samdong MgB_2 wire.

For the neutron measurements of the NIFS sample, the experiments were conducted under load control, with load steps at 0 N, 8 N, 16 N, 24 N, 32 N, 45 N, 60 N, 80 N, 100 N, 120 N, and 168 N at 11 points.

3. Results

3.1. *Diffraction Histogram by the ToF Method*

Figure 10 illustrates the load history of the NIFS wire, subjected to load levels of 0 N, 8 N, 16 N, 24 N, 32 N, 45 N, 60 N, 80 N, 100 N, 120 N, and 168 N. Each step allowed for approximately 1 h of irradiation time. The horizontal axis represents elapsed time in seconds, while the vertical axis denotes the load in Newtons, as measured by the load cell. With increasing load, stress relaxation occurs due to plasticity. Careful manual load control was employed to prevent excessive load reduction. Some experiments had extended durations due to beam stoppage times. Stress–strain diagrams were derived from strain and load data collected via a double extensometer and load cell. The data, acquired using a horizontal-type tensile frame, display slight non-linearity compared to data from the dedicated vertical tensile frame. Nonetheless, the gradients are generally consistent with those presented in Figure 7. Figure 11 displays the results, with the horizontal axis representing strain in percentage obtained from extensometers, and the vertical axis representing load in Newtons from the load cell. This stress–strain diagram includes data from unloading, revealing a residual strain of approximately 0.05% from the maximum strain of 0.216%.

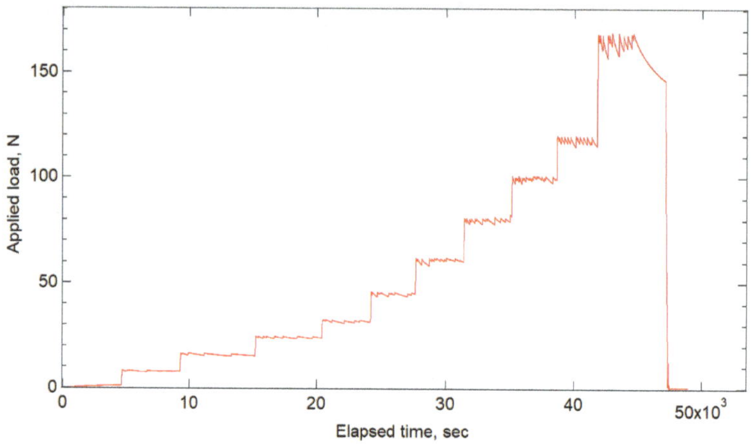

Figure 10. Loading history of NIFS.

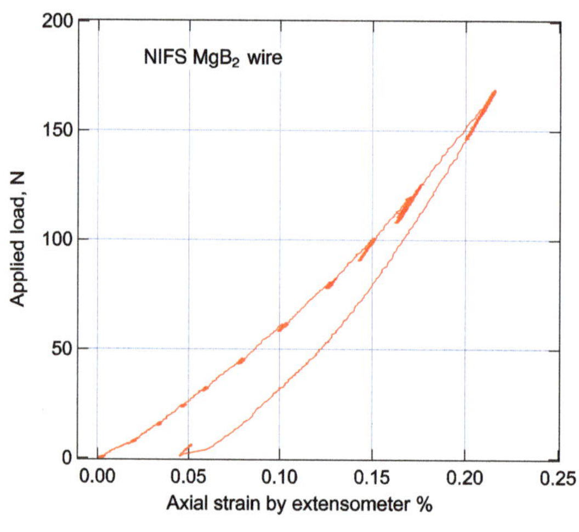

Figure 11. Stress–strain diagram of NIFS during measurement.

To generate histograms of Time-of-Flight (ToF) events, data from all North detector events in the 75° to 105° range were utilized. A 120 min test scan was conducted on any sample when the load was 0. The results for NIFS (^{11}B-enriched sample), Samdong, and Hypertech are presented in Figures 12–14, respectively. The vertical axis in these figures represents diffraction intensity, while the horizontal axis depicts ToF in microseconds, obtained from the axial lattice plane of the wire. This ToF value can be converted into common lattice spacing (in angstroms (Å)) by dividing it by approximately 15,000. The three-digit numbers in these figures denote the Miller indices of diffraction for each phase, determined through Rietveld analysis.

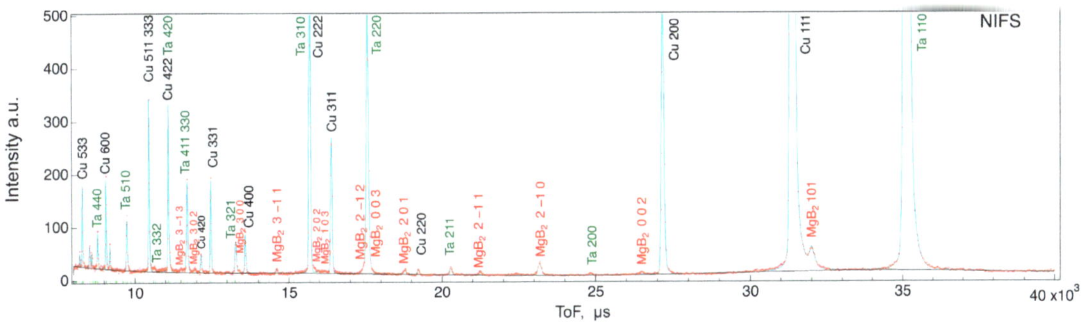

Figure 12. ToF histogram of NIFS ^{11}B-enriched MgB_2 wire.

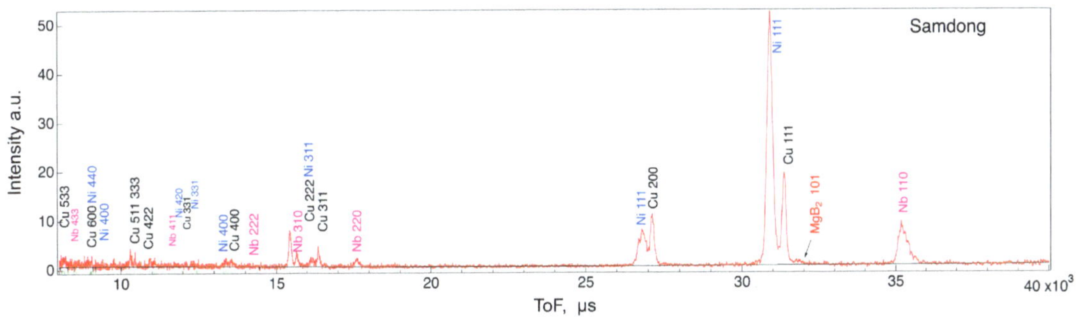

Figure 13. ToF histogram of Samdong MgB_2 wire.

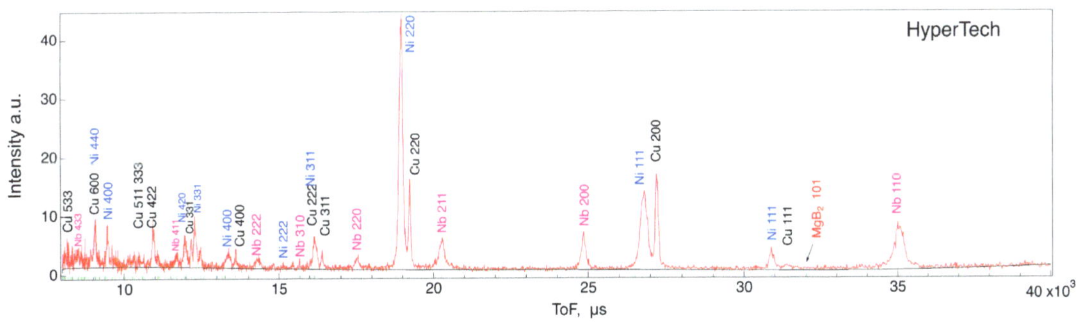

Figure 14. ToF histogram of Hypertech MgB_2 wire.

As shown in Figure 4, the wires are positioned at a 45° angle to the incident beam, resulting in North bank data that provide a histogram of lattice spacing (*d*) or ToF along the sample's axis direction. The results for the NIFS wire exhibit peaks corresponding to the included MgB_2, pure copper, and tantalum phases, with sufficient intensity for Rietveld analysis. For this analysis, we employed the Z-Rietveld software [22], capable of powder diffraction data analysis and Rietveld analysis, which was developed in Japan and includes instrument parameters specific to J-PARC. Rietveld analysis allows the determination of lattice constants for both refined '*a*' and '*c*' values, considering that MgB_2 exhibits a hexagonal crystal structure.

In general, crystals exhibit elastic anisotropy, meaning that some orientations are more deformable than others under the same stress, and different crystal planes possess distinct elastic constants. While the lattice spacing of crystals elongates in the axial direction during tensile tests, the rate of change varies from one lattice plane to another. Rietveld analysis permits the refinement of an average lattice constant, closely aligning with the mechanical properties. Although similar calculations can be performed using single peaks, the method employing Rietveld analysis is more prevalent for ToF instruments, as it enhances accuracy.

A few MgB_2 peaks were observed for the Samdong and Hypertech commercial wires in Figures 11 and 12. These wires are thought to contain a natural ratio of boron isotopes and about 20% ^{10}B, which has a large neutron absorption cross-section. As this experiment was carried out with a single wire, measures could be taken to increase the diffraction intensity by increasing the number of wires. However, assuming multi-twisted wires, such as superconducting wires, the diffraction intensity is likely to eventually decrease because of its own absorption problems. Materials with very large neutron absorption cross-sections, such as ^{10}B, do not prevent diffraction per se, but absorb diffraction themselves, resulting in a significant reduction in diffraction intensity.

3.2. Strain Measurement under Tensile Loadings

Given that the crystal system of MgB_2 is hexagonal, it is noteworthy that the lattice constants 'a' and 'b' are equivalent ($a = c$). However, the 'c' axis, representing the length of the hexagonal cross-section, remains an independent constant. Measurements were conducted at various load levels (0 N, 8 N, 16 N, 24 N, 32 N, 45 N, 60 N, 80 N, 100 N, 120 N, and 168 N), and the acquired data were subjected to Rietveld analysis for each load to derive the 'a' and 'c' values for MgB_2.

As for the sheath materials, Ta and Cu, they exhibit cubic systems, and the lattice constant '*a*' values were similarly determined for each load. Figure 15 depicts the relationship between the axial lattice constant of hexagonal MgB_2 and the applied load. The horizontal axis in the figure represents the load in Newtons, obtained from the load cell, while the vertical axis displays the '*a*' or '*c*' axis length refined via Rietveld analysis in angstroms (Å). Each marker on the graph is accompanied by vertical lines representing error bars from the Rietveld analysis. It is worth noting that the error bars for the '*c*' axis tend to be larger due to the relatively lower number of diffractions linked to '*l*' across the numerous diffraction planes '*h k l*' obtained.

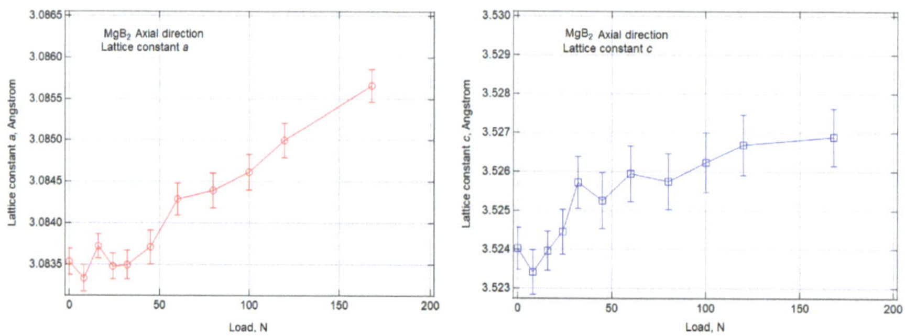

(a) (b)

Figure 15. Relation between the results of axial lattice constant changes of hexagonal MgB_2 and applied load calculated by Rietveld analysis: (**a**) lattice constant *a*; (**b**) lattice constant *c*.

From Figure 15a, we observe that the 'a' axis demonstrates a linear variation within the range from 80 N to 168 N. In contrast, Figure 15b reveals that the 'c' axis exhibits a linear change up to 32 N, after which the change becomes more gradual. This issue should be considered in conjunction with other strain-related factors after converting 'd' to lattice strain, a discussion which will be addressed in the subsequent section.

Figure 16 depicts the variations in the axial lattice constants of the sheath materials, tantalum (Ta) and copper (Cu). In this figure, the horizontal axis represents the load in Newtons, as measured by the load cell, while the vertical axis displays the axial lattice constant 'a' of cubic Cu or Ta, refined through Rietveld analysis, in angstroms (Å).

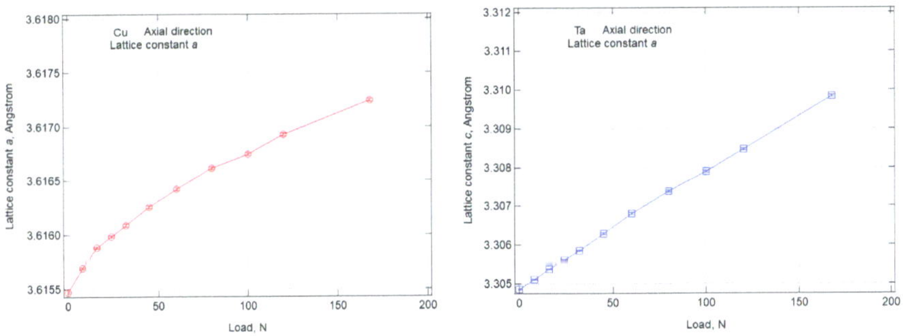

(a) (b)

Figure 16. Relation between the results of the axial lattice constant changes of sheath and applied load calculated by Rietveld analysis: (**a**) Cu phase *a*; (**b**) Ta phase *a*.

Figure 16a illustrates the results for Cu, where the curve maintains linearity up to the second data point, 16 N. Beyond this point, the curve exhibits typical plastic deformation. Cu is likely to have pre-existing tensile residual stresses due to its inherently low initial strength and CTE (coefficient of thermal expansion) mismatch, making it prone to yielding even under low stresses.

On the other hand, Ta, as shown in Figure 16b, boasts the highest strength among the constituent materials and exhibits a consistent linear behavior from the beginning to the end. This linear behavior of tantalum reaffirms that the loads and strains are acting upon the wire as expected.

Similarly, the changes in the lattice spacing in the lateral direction, obtained from the South detector, are summarized in Figure 17. Once again, the horizontal axis represents the load in Newtons, measured by the load cell, while the vertical axis displays the '*a*' or '*c*' axis length refined through Rietveld analysis in angstroms (Å). In this direction, the influence of Poisson's ratio leads to a shrinkage deformation, causing the lattice spacing to decrease as the load increases. This behavior appears to follow a linear trend from 0 to approximately 80 N.

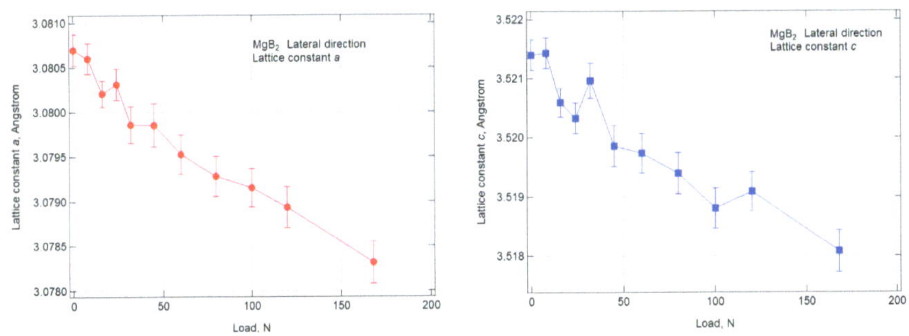

(a) (b)

Figure 17. Relation between the results of the lateral lattice constant changes of hexagonal MgB_2 and applied load calculated by Rietveld analysis: (**a**) lattice constant *a*; (**b**) lattice constant *c*.

A summarized overview of the changes in axial and lateral lattice constants is presented in Figure 18 for lattice constant '*a*' and Figure 19 for lattice constant '*c*'. In both of these figures, the horizontal axis represents the load in Newtons (N), as measured by the load cell, while the vertical axis displays the length of the '*a*' or '*c*' axis refined through Rietveld analysis, expressed in angstroms (Å).

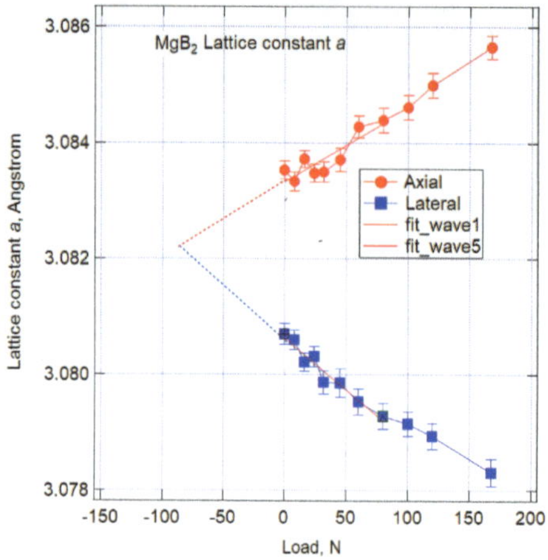

Figure 18. Relation between the results of lattice constant changes of hexagonal MgB_2 '*a*' and applied load (axial and lateral).

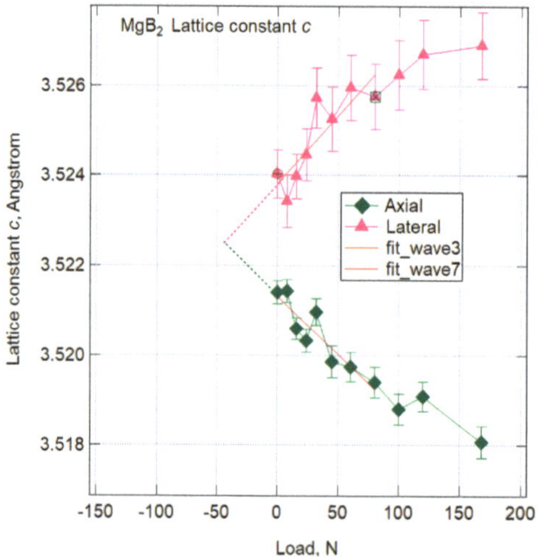

Figure 19. Relation between the results of lattice constant changes of hexagonal MgB_2 '*c*' and applied load (axial and lateral).

During such tensile tests, it is common for Poisson's ratio to be approximately 0.3, signifying that the lateral strain is about one-third of the axial strain. The slopes of the graphs in Figures 18 and 19, corresponding to lattice constant '*a*' and '*c*', respectively, exhibit no significant difference in slope, although there are both positive and negative variations. This suggests that strain in the lateral direction is transmitted predominantly through compression, while strain in the tensile direction may experience relaxation through processes such as cracking, especially in the initial tensile phase.

It is important to take into account residual stresses. For this study, a filament d_0 sample, which serves as the strain-free reference for residual strain or stress, was not prepared. This wire underwent heat treatment in the final stage of the manufacturing process, which likely resulted in residual strain or stress due to the effects of CTE (coefficient of thermal expansion) mismatch. While simplified, residual stresses can be estimated by comparing axial and lateral lattice constants. In the absence of residual strain, both lattice constants start from the same point, with axial values increasing and lateral values decreasing, leading to an increasing difference between them.

In both diagrams, lines of best-fit approximation derived from the initial eight loading points are displayed. In this case, both figures indicate that the axial lattice plane spacing is larger, and the lateral values are smaller when examining the 0 N points. This observation suggests that the MgB_2 filament experiences residual tensile strain or stress at room temperature, albeit in a qualitative sense. By extending the previously mentioned straight line to the compression side, one can simply estimate the point where the lattice plane spacing coincides, representing minimal strain. The estimated loads for this point were approximately −85 N for the *a*-axis and −45 N for the *c*-axis.

4. Discussion

The Time-of-Flight (ToF) data were processed by applying conversion parameters to obtain the lattice spacing (*d*). Subsequently, we transformed these values into lattice strain for each crystal phase using Equation (2). When no load was applied (load = 0), the strain was set to d_0. Figure 20 illustrates the correlation between the lattice strain results of each phase and the applied load using the ToF method of Takumi. The horizontal axis of this figure represents the load in Newtons obtained from the load cell, while the vertical axis represents the elastic strain of the crystal, expressed in dimensionless units, derived from the lattice constants of each phase (MgB_2, Cu, and Ta). Each marker on the graph is accompanied by a vertical error bar, which represents the error from Rietveld analysis converted into strain. Notably, the results for Ta exhibited excellent linearity across all regions, allowing us to draw an approximate straight line using the least-squares method specifically for the Ta results. In the case of Ta, the diagram displays a consistently linear relationship between strain and load, with minimal error bars. The fact that such a strong linear relationship exists for a metallic element like tantalum indicates the accurate loading of the wire.

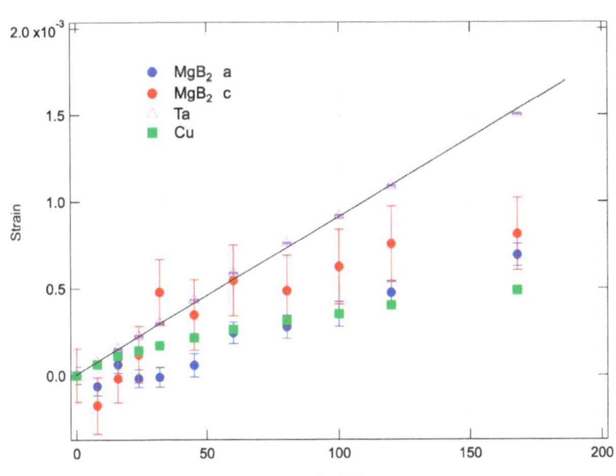

Figure 20. Relation between the results of lattice strain of each phase and applied load.

In strain measurements employing diffraction techniques, only elastic strain is ascertained, and plastic strain, as observed during yielding, remains undetected. Consequently, we observe that Cu begins to yield at approximately 16 N, and strain sharing increases with work hardening. The slope of Ta's behavior is contrasted with that of MgB_2, with the assumption that Ta's slope equals the slope of linear elastic engineering strain. Given this assumption, it appears that all phases except Cu exhibit linear changes up to 60 N. While the 'a' axis of MgB_2 generally displayed linear changes in the lattice constant data presented in Figure 15a, it is reasonable to infer that the lattice strain shows a decreasing gradient from 60 N when converted. A load of 60 N corresponds to a stress of 67 MPa on the stress–strain diagram in Figure 7, where the strain is approximately 0.09%. The disparity between the lattice parameters in Figures 18 and 19 implies that the filaments in this wire may have already experienced tensile residual stress or strain, making them susceptible to failure at relatively low stress levels. Notably, the 'c' axis values of MgB_2 at 168 N reveal distinct strain relaxation, suggesting that filament rupture may be occurring. Since diffraction only allows the observation of elastic strain, any cracking occurring at grain boundaries or within grains can be detected as such strain or stress relaxation.

We have conducted similar experiments with Nb_3Sn [11], BSCCO [12], and REBCO [9,10,23] superconducting wires to determine the filament breakage strain. However, experiments with MgB_2 have proven challenging, and this marks the first measurement of such strain experiments under load. If experiments involving tensile loading at low temperatures can be conducted, and strain-free standard d_0 samples are obtained, we can gain insights into the residual strain at actual temperatures and the breakage strain of filaments. Since we have successfully conducted strain measurements at room temperature, future experiments at lower temperatures will be the focus of our research. The low-temperature tensile testing frame used in this study can be cooled down to approximately 10 K, and as cryogenic tensile frames are already available, we hope to employ the findings from this study to unravel the unresolved mystery of residual strain in MgB_2 wire. Another significant objective of this research is to establish the relationship between tensile load, critical current, and actual strain on the filament at low temperatures. This will pave the way for guidelines on how to control residual strain through the optimization of sheath materials and the strengthening of the filaments themselves.

5. Conclusions

(1) Neutron scattering experiments conducted on ^{11}B-enriched MgB_2 wire have successfully yielded abundant diffraction peaks specific to MgB_2. The subsequent Rietveld analysis of each phase enabled the fine-tuning of lattice constants, incorporating nearly all available peaks. The ratio of the change in lattice constant to the load in the axial and lateral directions was close, indicating that the axial strain may be relaxing.

(2) This groundbreaking analysis has unveiled previously unreported variations in the load and strain experienced by MgB_2 wires subjected to tensile stress at room temperature. The discrepancy in lattice constants between the axial and lateral directions strongly implies the presence of tensile residual stress or strain at room temperature.

(3) The proportionate change in lattice constant relative to the applied load in both axial and lateral directions exhibits remarkable similarity, suggesting the potential relaxation of axial strain.

(4) Notably, in experiments involving two commercially available, conventional MgB_2 wire types, neutron scattering experiments were conducted for a duration of 7200 s. However, no discernible MgB_2 peaks suitable for analysis were obtained. This observation strongly indicates the formidable challenge of achieving MgB_2 diffraction in neutron scattering experiments when dealing with boron-containing natural isotopes with their inherent ratios.

Author Contributions: Conceptualization, S.M., K.O. and Y.H.; formal analysis, S.M., S.H., H.T. and T.K.; data curation, S.M., H.T. and T.K.; writing—original draft preparation, S.M.; writing—review and editing, S.M., K.O. and Y.H. All authors have read and agreed to the published version of the manuscript.

Funding: This study was performed with the support and under the auspices of the NIFS Collaboration Re-search program (NIFS18KECF023, NIFS21KECF031).

Data Availability Statement: The data presented in this study are available on request from the corresponding author.

Acknowledgments: The neutron experiment at BL19 Takumi of the Materials and Life Science Experimental Facility of the J-PARC was performed under user program 2019A0178.

Conflicts of Interest: The authors declare no conflict of interest.

References

1. Nagamatsu, J.; Nakagawa, N.; Muranaka, T.; Zenitani, Y.; Akimitsu, J. Superconductivity at 39 K in magnesium diboride. *Nature* **2001**, *410*, 63–64. [CrossRef] [PubMed]
2. Ballarino, A.; Flükiger, R. Status of MgB_2 wire and cable applications in Europe. *J. Phys. Conf. Ser.* **2017**, *871*, 012098. [CrossRef]
3. Marino, I.; Pujana, A.; Sarmiento, G.; Sanz, S.; Merino, J.M.; Tropeano, M.; Sun, J.; Canosa, T. Lightweight MgB_2 superconducting 10 MW wind generator. *Supercond. Sci. Technol.* **2016**, *29*, 024005. [CrossRef]
4. Tanaka, H.; Suzuki, T.; Kodama, M.; Koga, T.; Watanabe, H.; Yamamoto, A.; Michizono, S. Performance of MgB_2 superconductor developed for high-efficiency klystron applications. *IEEE Trans. Appl. Supercond.* **2020**, *29*, 6200105. [CrossRef]
5. Choi, Y.; Park, D.; Li, Y.; Tanaka, H.; Lee, E.; Bascuñán, J.; Iwasa, Y. Persistent-mode operation and magnetization behavior of a solid nitrogen-cooled MgB_2 small-scale test coil towards a tabletop 1.5-T osteoporosis MRI. *Supercond. Sci. Technol.* **2020**, *33*, 1018. [CrossRef]
6. Sykes, A.; Costley, A.E.; Windsor, C.G.; Asunta, O.; Brittles, G.; Buxton, P.; Chuyanov, V.; Connor, J.W.; Gryaznevich, M.P.; Huang, B.; et al. Compact fusion energy based on the spherical tokamak. *Nucl. Fusion* **2018**, *58*, 016039. [CrossRef]
7. Leslie, M.; Ito, T.; Aizawa, K.; Arima, H. Start-Ups Seek to Accelerate Path to Nuclear Fusion. *Engineering* **2021**, *8*, 6–8. [CrossRef]
8. Ekin, W.J. Chapter "Strain Effects in Superconducting Compounds". In *Advances in Cryogenic Engineering Materials*; Plenum Press: New York, NY, USA, 1984; Volume 30.
9. Osamura, K.; Machiya, S.; Tsuchiya, Y.; Suzuki, H. Internal Strain and Mechanical Properties at Low Temperatures of Surround Cu Stabilized YBCO coated conductor. *IEEE Trans. Appl. Supercond.* **2010**, *20*, 1532–1536. [CrossRef]
10. Osamura, K.; Machiya, S.; Tsuchiya, Y.; Suzuki, H. Force free strain exerted on a YBCO layer at 77 K in surround Cu stabilized YBCO coated conductors. *Supercond. Sci. Technol.* **2010**, *23*, 045020–045026. [CrossRef]
11. Oguro, H.; Awaji, S.; Nishijima, G.; Takahashi, K.; Watanabe, K.; Machiya, S.; Suzuki, H.; Tsuchiya, Y.; Osamura, K. Residual strain measurement using neutron diffraction for practical Nb_3Sn wires under a tensile load. *Supercond. Sci. Technol.* **2010**, *23*, 025034. [CrossRef]
12. Osamura, K.; Machiya, S.; Ochiai, S.; Osabe, G.; Yamazaki, K.; Fujikami, J. High strength/high strain tolerance DI-BSCCO tapes by means of pre-tensioned lamination technique. *IEEE Trans. Appl. Supercond.* **2013**, *23*, 6400504. [CrossRef]
13. Hishinuma, Y.; Kikuchi, A.; Shimada, Y.; Kashiwai, T.; Hata, S.; Yamada, S.; Muroga, T.; Sagara, A. Development of MgB_2 superconducting wire for the low activation superconducting magnet system operated around core D-T plasma. *Fusion Eng. Des.* **2015**, *98–99*, 1076–1080. [CrossRef]
14. Hishinuma, Y.; Shimada, Y.; Hata, S.; Tanaka, T.; Kikuchi, A. Superconducting Properties and Microstructure of In-situ Cu Addition Low Activation MgB_2 Multifilamentary Wires Using Different Boron-11 Isotope Powders. *Teion Kogaku* **2022**, *57*, 32–38. (In Japanese)
15. Cheng, F.; Liu, Y.; Ma, Z.; Li, H.; Shahriar, M.; Hossain, A. Superior critical current density obtained in $Mg_{11}B_2$ low activation superconductor by using reactive amorphous ^{11}B and optimizing sintering temperature. *J. Alloys Compd.* **2015**, *650*, 508–513. [CrossRef]
16. Mooring, F.P.; Monahan, J.E.; Huddleston, C.M. Neutron cross sections of the boron isotopes for energies between 10 and 500 keV. *Nucl. Phys.* **1966**, *82*, 16–32. [CrossRef]
17. Harjo, S.; Ito, T.; Aizawa, K.; Arima, H. Current Status of Engineering Materials Diffractometer at J-PARC. *Mater. Sci. Forum* **2013**, *681*, 443–448. [CrossRef]
18. Kikuchi, A.; Yoshida, Y.; Iijima, Y.; Banno, N.; Takeuchi, T.; Inoue, K. The synthesis of MgB_2 superconductor using Mg_2Cu as a starting material. *Supercond. Sci. Technol.* **2004**, *17*, 781–785. [CrossRef]
19. Jin, X.; Nakamoto, T.; Harjo, S.; Hemmi, T.; Umeno, T.; Ogitsu, T.; Yamamoto, A.; Sugano, M.; Aizawa, K.; Abe, J.; et al. Development of a cryogenic load frame for the neutron diffractometer at Takumi in Japan Proton Accelerator Research Complex. *Rev. Sci. Instrum.* **2013**, *84*, 063106. [CrossRef]
20. Nyilas, A. Strain sensing systems tailored for tensile measurement of fragile wires. *Supercond. Sci. Technol.* **2005**, *18*, S409. [CrossRef]
21. *IEC 61788 19: 2013*; Mechanical properties measurement-Room temperature tensile test of reacted Nb_3Sn composite superconductors.

22. Oishi, R.; Yonemura, M.; Nishimaki, Y.; Torii, S.; Hoshikawa, A.; Ishigaki, T.; Morishima, T.; Mori, K.; Kamiyama, T. Rietveld analysis software for J-PARC. *Nucl. Instrum. Methods Phys. Res. Sect. A* **2009**, *600*, 94–96. [CrossRef]
23. Sugano, M.; Machiya, S.; Osamura, K.; Adachi, H.; Sato, M.; Semerad, R.; Prussei, W. The direct evaluation of the internal strain of biaxially textured YBCO film in a coated conductor using synchrotron radiation. *Supercond. Sci. Technol.* **2009**, *22*, 015002. [CrossRef]

quantum beam science

MDPI

Article

Neutron Stress Measurement of W/Ti Composite in Cryogenic Temperatures Using Time-of-Flight Method

Masayuki Nishida [1,*], Stefanus Harjo [2], Takuro Kawasaki [2], Takayuki Yamashita [3] and Wu Gong [2]

1 Department of Mechanical Engineering, Kobe City College of Technology, 8-3 Gakuenhigashi-Machi, Nishi-Ku, Kobe 651-2194, Japan
2 J-PARC Center, Japan Atomic Energy Agency, Tokai-Mura, Naka-Gun, Ibaraki 319-1195, Japan
3 Joining and Welding Research Institute, Osaka University, 11-1 Mihogaoka, Ibaraki, Osaka 567-0047, Japan
* Correspondence: nishida@kobe-kosen.ac.jp; Tel.: +81-078-795-3212

Abstract: In this study, the thermal stress alterations generated in a tungsten fiber reinforced titanium composite (W/Ti composite) were evaluated by the neutron stress measurement method at cryogenic temperatures. The W/Ti composite thermal loads were repeated from room temperature to the cryogenic temperature (10 K), and alterations in thermal residual stress were evaluated using the neutron in situ stress measurement method. In this measurement, the stress alterations in the titanium matrix and the tungsten fibers were measured. This measurement was carried out by TAKUMI (MLF-BL19) of J-PARC, a neutron research facility in the Japan Atomic Agency. The measurement method of TAKUMI is the time-of-flight (TOF) method. Owing to this measurement method, the measurement time was significantly shortened compared to the angle-dispersion type measurement by a diffractometer. As a result of the measurement, large compressive stresses of about 1 GPa were generated in the tungsten fibers, and tensile stresses of about 100 MPa existed in the titanium matrix. The thermal stresses due to the temperature change between room temperature and cryogenic temperature is caused by the difference of thermal expansions between the tungsten fibers and the titanium matrix, and these stress values can be approximated by a simple elastic theory equation.

Keywords: composite materials; thermal stresses; neutron diffraction; time-of-flight; in situ stress measurement

1. Introduction

Fiber-reinforced metal composites (FRM) are composite materials which have a high specific strength and specific stiffness, as well as excellent wear resistance and environmental resistance. Compared to carbon fiber-reinforced plastic (CFRP), FRM is expected to be used in special environments. For example, in the case of long-term use in outer space developments, many problems exist, such as high vacuum, strong ultraviolet rays, temperature difference, radiation, and the presence of primitive oxygen [1]. The use of CFRP is difficult for these reasons, and FRM is promising in this field [1–3]. In addition, as a practical example, it has been used for parts in conditions of high temperature and/or high pressure such as piston heads of automobile engines [4,5].

On the other hand, the metal matrix of FRM itself has high strength and good weldability. For this reason, when compared to the view point of the joint problems, which is the weakest point of CFRP, FRM is more advantageous than CFRP because it is easier to joint and weld large structures and parts for high-pressure environments [6–8]. In this way, the feature of FRM is that it is assumed to be used in extreme environments.

In recent years, the development of materials with high strength for use in cryogenic environments such as liquefied natural gas (111 K) tanks and liquid hydrogen (23.6 K) tanks has progressed. In particular, in order to reduce fuel transportation costs, there is an urgent need to increase the strength and reduce the weight of cryogenic storage tanks. The application of composite materials is promising for high strength and weight reduction. However, in the case of CFRPs, there are many problems such as low-temperature embrittlement in the polymer matrix and generation of cracks [1]. On the other hand, FRMs are presumed to be advantageous for use at extremely low temperatures, because aluminum and titanium are often used as metal matrix without causing low-temperature embrittlement. It is also important that FRM is advantageous for jointing and welding, which are essential in the production of storage tanks.

Comparing the mechanical properties of titanium and aluminum at cryogenic temperatures [9,10], the specific strength of titanium is three times greater than that of aluminum, and the thermal conductivity of titanium is three orders of magnitude lower than that of aluminum. Furthermore, the coefficient of thermal expansion is about 30% that of aluminum, and titanium has higher corrosion resistance (especially seawater corrosion resistance) than aluminum. These properties make titanium superior to aluminum as a structural material in cryogenic and/or corrosive environments. Furthermore, titanium can be further enhanced by alloying it, such as Ti-6Al-4V. On the other hand, titanium has excellent properties as a structural material used at low temperatures; however, titanium is used as a single element rather than as a composite material at present. If titanium is reinforced with high-strength fiber, further improvement in performance can be expected; however, there are almost no research reports on titanium fiber-reinforced composites.

In this study, the W/Ti composite material was manufactured by reinforcing the titanium matrix with tungsten fibers. This W/Ti composite material was fabricated by a simple continuous spot-welding method that does not require special equipment such as a vacuum chamber or high-temperature furnace [11]. There is an important point here. Titanium has extremely high metal activity at high temperatures, so if other metals are added to high temperature molten titanium, it will easily react and the metals will diffuse into the titanium and disappear. Therefore, it is impossible to manufacture titanium base metal fiber-reinforced material using a casting method. On the other hand, the spot-welding method employed in this study is a method of joining the titanium matrix with tungsten fibers without completely melting. On the contrary, the semi-molten titanium diffuses slightly into tungsten fibers, and the bonding of the titanium matrix and the tungsten fibers becomes a metallurgical bond. As a result, the bonding between the titanium matrix and the tungsten fibers becomes strong.

In the future, the W/Ti composite investigated in this study may be used for tanks for cryogenic materials such as liquefied natural gas and liquid hydrogen, as well as for liquid hydrogen in addition to liquid oxygen tanks for rocket fuel for spaceships. However, the fundamental mechanical properties of W/Ti composites as structural materials have not yet been investigated in detail.

The purpose of this study is to evaluate thermal residual stresses generated inside the W/Ti composite at cryogenic environments. Commonly, in any fiber-reinforced composite, there is a large difference in thermal expansion coefficients between the matrix and the fiber. This mismatch causes the thermo-induce residual stresses in every composite material [12,13]. These thermal residual stresses are big problems which can never be avoided in all of the fiber-reinforced composite materials. Among such composite materials, the W/Ti composite manufactured in this study is expected to reduce thermal residual stresses because the thermal expansion coefficient of titanium matrix is smaller than that of aluminum and steel at cryogenic temperatures.

The existence of such thermal residual stresses is a very important parameter to consider in the design of strengths. Usually, these thermal residual stresses are superimposed with the initial residual stress generated depending on the thermal history during the manufacturing of the composite material and the thermal residual stress due to the temperature alterations in the environment in which the composite material is used. In this study, both the initial residual stress generated from the thermal history during fabrication and the thermal stresses generated from the environmental temperature alterations of the composite material were measured.

Since these thermal stresses are generated inside the W/Ti composite, it is necessary to nondestructively measure the internal stress while applying cryogenic temperature alterations to the W/Ti composite. The neutron stress measurement technique is useful for such nondestructive in situ measurements inside metals. Nondestructive internal stress evaluation is impossible using any other method. An effective method has been developed for analyzing the thermal stress in the internal position of composite materials using the neutron stress measurement technique [14,15]. This measurement method is essentially based on Hooke's law with lattice spacing alterations.

At this time, the measurement was carried out by TAKUMI (MLF-BL19) of J-PARC, a neutron research facility at the Japan Atomic Agency. The measurement method of TAKUMI is the time-of-flight (TOF) method [16]. Owing to this measurement method, the measurement time was significantly shortened compared to the angle-dispersion type measurement by a diffractometer. Furthermore, since the measurement range of diffraction profiles became wide in one time measurement, it is possible to evaluate stresses on multiple diffraction planes. In this study, the W/Ti composite was set in a cryostat cooling system mounted on the sample table of TAKUMI to generate the cryogenic temperature states. The W/Ti composite thermal loads were repeated from room temperature (279 K) to the cryogenic temperature (10 K), and alterations in thermal residual stresses were evaluated by the neutron in situ stress measurement method. The stress alterations in the titanium matrix and the tungsten fibers were measured at each temperature.

2. Materials and Methods

2.1. Preparation of Fiber Reinforced Material

The W/Ti composite was produced for this investigation. In this W/Ti composite, 99.99% purity tungsten fiber with 100 μm diameter and 99.9% purity industrial titanium plates of thickness 0.5 mm and 0.2 mm were used for the fiber phase and the matrix phase, respectively. In this study, the W/Ti composite was produced by the continuous spot welding method. This manufacturing method uses only a simple spot weld, and it does not need a vacuum chamber or a high temperature furnace such as existing common methods. The arranged tungsten fibers were held between titanium plates and fixed by spot welding. Furthermore, all surface areas were spot-welded continuously while moving these materials. This continuous spot-welding method supplements the small localized welds of spot welding when joining is required over the entire area of W/Ti composites. Finally, the tungsten fiber and the titanium plate were joined together to form the whole W/Ti composite. The coverage, which was a rate of welding area to the whole plate surface, became 150% for the W/Ti composite in this manufacturing.

Figure 1a–c show the schematic diagram of the manufacturing method for the W/Ti composite by continuous spot welding. Owing to a change in the thickness of the titanium plate and/or the spacing of the tungsten fiber arrangements, the volume ratio of the tungsten fibers could be freely adjusted in the W/Ti composite.

To prepare for this study, Figure 1a shows a titanium plate (0.5 mm thick) with tungsten fibers wound at regular intervals and sandwiched between other titanium plates (0.2 mm thick). As shown in Figure 1b, seven layers of these materials prepared in Figure 1a are stacked up and spot-welded. Figure 1c shows the overlapping condition of the welded part in the continuous spot-welding method, and the welding path on the surface of the W/Ti composite.

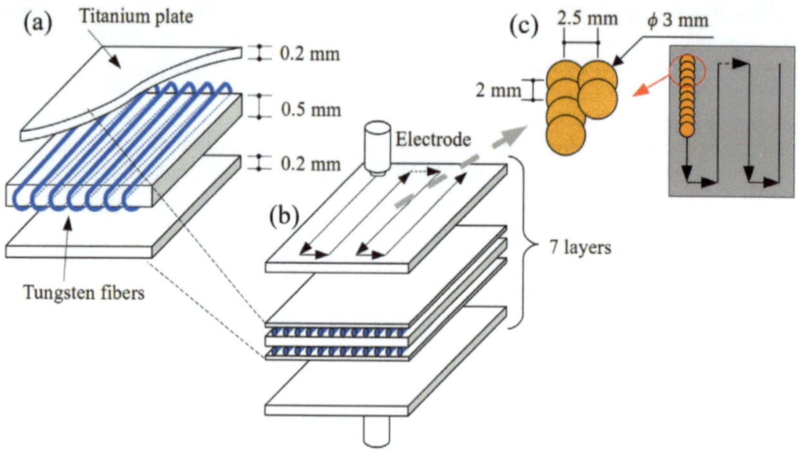

Figure 1. Schematic diagram of (**a**) 0.5 mm thick titanium plate with tungsten fibers wound at regular intervals and sandwiched between other 0.2 mm thick titanium plates; (**b**) seven layers of these materials prepared in (**a**) are stacked up and spot-welded; (**c**) the overlapping condition of the welded part in the continuous spot-welding method, and the welding path on the surface of W/Ti composite.

Figure 2a,b show SEM photographs of the W/Ti composite. Figure 2a is the SEM photograph of the cross-section of the W/Ti composite. The distances between the lines of tungsten fibers are about 0.5 mm and 0.2 mm. These arrangements depend upon the arrangement space of tungsten fibers and the thickness of titanium plates. Figure 2b shows the results of SEM component analysis focusing on boundary between the titanium matrix and the tungsten fibers.

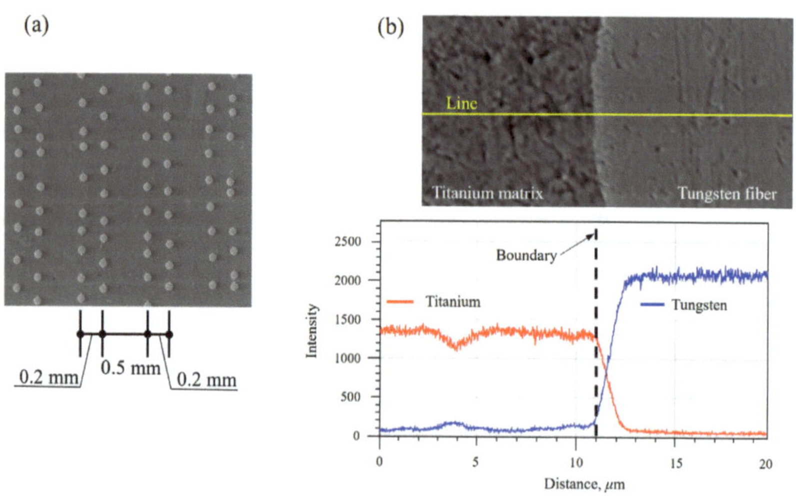

Figure 2. SEM photographs of the W/Ti composite: (**a**) a photograph of the cross-section of the W/Ti composite; (**b**) SEM component analysis focusing on the boundary between the titanium matrix and the tungsten fibers.

The past measurement results confirmed that the titanium matrix diffuses into the tungsten fiber depending on the welding conditions such as current value, sample temperature, and coverage during continuous spot welding. In the SEM observation results in this study, it could be confirmed that the titanium matrix diffuses to the tungsten fiber side slightly beyond the boundary between titanium and tungsten. Furthermore, no oxide layer was observed between the titanium matrix and the tungsten fibers. This result indicates that the tungsten fibers and the titanium matrix are bonded metallurgically, rather than the titanium matrix simply covering the tungsten fibers.

The conditions of the spot welding are shown in Table 1. The final dimension of the W/Ti composite is 115 mm × 30 mm, with thickness of 7.0 mm. From this W/Ti composite, the measurement sample for the in situ thermal residual stress measurement of TAKUMI was cut off by 12 mm × 12 mm, with a thickness of 7.0 mm. The volume fraction of the tungsten fiber was about 5% in this sample.

Table 1. Conditions of spot welding.

Welding voltage (V) & current (A)	200, 8.5
Welding pressure (kN)	19
Holding time (msec.)	200
Diameter of electrode (mm)	11

2.2. In Situ Thermal Stress Measurement

The W/Ti composite evaluated in this investigation is schematically shown in Figure 3a. The x_1 axis is defined as parallel to the longitudinal direction of the tungsten fibers. The x_2 and x_3 axes are normal to the fiber direction. When stresses are calculated by Hooke's equation, the strains in the three directions of x_1, x_2, and x_3 are required. In this measurement, it was assumed that stresses and strains were almost equal in the x_2 and x_3 directions. Therefore, only two directions in the x_1 axis and the x_2 axis were measured by the neutron diffraction.

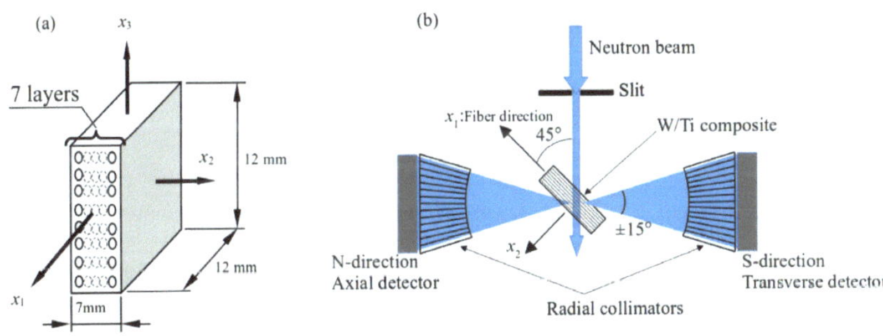

Figure 3. Schematic diagram of (**a**) coordinate system of W/Ti composite, and (**b**) sample setting and neutron measurement image.

Figure 3b shows the setting condition of the W/Ti composite on the sample table of TAKUMI. Because TAKUMI has two detectors, it is possible to measure diffraction profiles in two directions with one time measurement. These directions in Figure 3b are recorded as the north (N) and south (S) directions in the measurement file names from the geographical construction position of TAKUMI(BL19). The sample of the W/Ti composite is set at a position tilted at 45° from the incident direction of the neutron beam. In the measurement results, the N-direction becomes measurement data of the longitudinal direction of the tungsten fibers (x_1 direction), and the S-direction is the fiber normal direction ($x_2 = x_3$).

In the actual measurement, when the target temperature is reached, the TAKUMI sample stage is moved to the measurement position of three samples by sliding with the cryostat on it. Therefore, three time measurements are made for one temperature.

Figure 4a–d show photographs of three samples prepared for the in situ stress measurements. Figure 4a is a d_0 sample of the titanium. This sample was manufactured by the spot weld of titanium plates without tungsten fibers. Figure 4b is the W/Ti composite material evaluated in this study. Figure 4c is a tungsten sample for d_0 measurement, in which tungsten fiber is loosely wrapped around a titanium plate. These three samples were pasted to a copper plate by glue and tape. Figure 4d shows the sample setting condition on the cryostat cooling head. In this photograph, three samples were fixed with a thin white tape, so that they did not fall off from the copper plate due to shrinkage of cooling cycles. This white tape is a water leak prevention tape used for water pipe leaks. These measurement techniques are used by JAEA staff who assist the measurement on site. Such know-how is very important for actual measurement, and the authors were provided with full support by JAEA staff.

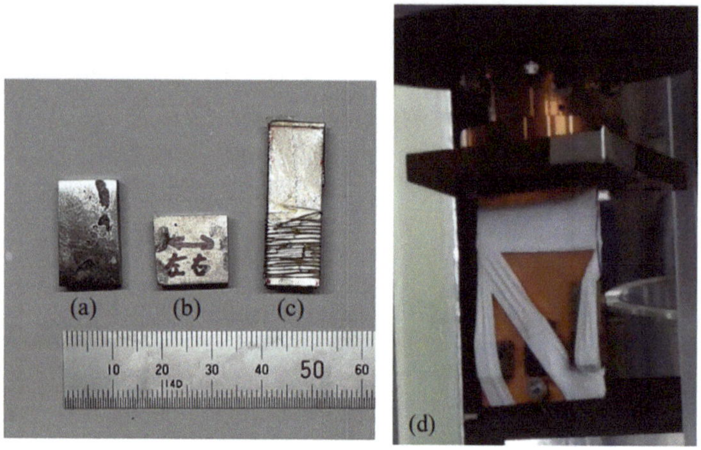

Figure 4. Photograph of the sample condition for the neutron measurement: (**a**) d_0 sample of the titanium manufactured by the spot weld of titanium plates without tungsten fibers; (**b**) the W/Ti composite material; (**c**) tungsten sample for d_0 measurement, where tungsten fiber is loosely wrapped around a titanium plate; (**d**) the sample setting condition on the cryostat cooling head. Three samples were fixed with a thin white water leak prevention tape.

In this measurement, a sample of tungsten fibers wrapped around titanium plates was prepared for d_0 measurement. However, the measurement accuracy of the measurement results deteriorated because the peaks of titanium and tungsten overlapped. In neutron stress measurement, d_0 measurement in a stress-free state is very important. As an improvement method, when measuring d_0 of several materials, it is necessary to measure the samples separately for d_0 measurement without combining them.

In this study, stress alterations due to low-temperature cycling were measured by the in situ neutron stress measurement technique. During the measurement of TAKUMI, the recording of diffraction profile data using the time-of-flight method, sample movement, and temperature control of the cryostat system were centrally managed by the main computer system.

Figure 5 shows a schematic diagram of the temperature vs. the time program for the in situ stress measurement of the W/Ti composite by TAKUMI. Measurement temperatures are seven points of 270 K, 250 K, 200 K, 150 K, 100 K, 50 K, and 10 K. The rate of temperature change is about 3.5 K/min in the cool down and heat up stages. When the temperature came to a target position, it was held for about 15 min in order to stabilize; after that, the stress measurement started in every case. The temperature during measurement was held at ±0.1 °C. One cycle consists of the cool down stage and the heat up stage, and three cycles were repeated in this measurement.

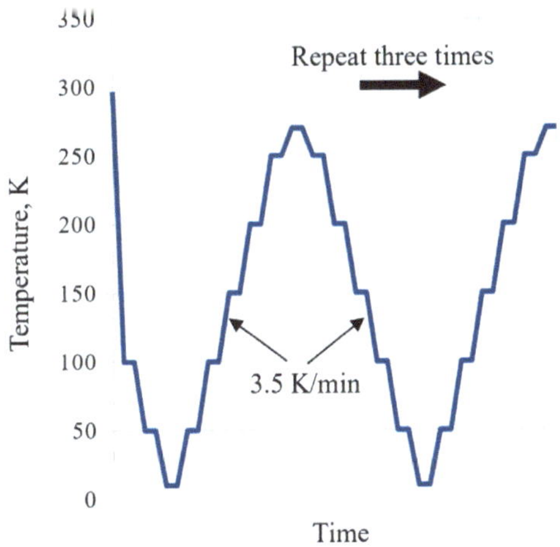

Figure 5. Schematic diagram of the temperature vs. time program for the in situ stress measurement of the W/Ti composite.

The thermal residual stresses σ_1 parallel to the longitudinal fiber direction and σ_2 normal to the longitudinal fiber direction were measured using Hooke's Equation (1).

$$\begin{cases} \sigma_1 = \frac{E}{(1+\nu)}\left\{\varepsilon_1 + \frac{\nu}{(1-2\nu)}(\varepsilon_1+\varepsilon_2+\varepsilon_3)\right\} \\ \sigma_2 = \frac{E}{(1+\nu)}\left\{\varepsilon_2 + \frac{\nu}{(1-2\nu)}(\varepsilon_1+\varepsilon_2+\varepsilon_3)\right\} \\ \sigma_3 = \frac{E}{(1+\nu)}\left\{\varepsilon_3 + \frac{\nu}{(1-2\nu)}(\varepsilon_1+\varepsilon_2+\varepsilon_3)\right\} \end{cases} \quad (1)$$

Table 2 shows the conditions of the in-situ neutron stress measurement. In this measurement, Young's modulus E and Poisson's ratio ν of the titanium matrix depend on the *hkl* diffraction plane. These material parameters were calculated from the Kroener model in the home page of "The Committee on X-ray Study on Mechanical Behavior of Materials, Japan" [17].

Table 2. Conditions of neutron stress measurement by TAKUMI.

MLF beam power	600 kW			
Measurement material	W/Ti composite			
Slit system	Incident slit: 5 × 10 mm Detected 90° A, B banks			
Measurement method	Time of Flight (TOF)			
Measurement time	600 sec./profile			
Young's modulus E (GPa) Poisson's ratio ν	Tungsten:			E: 388.69, ν: 0.2833
	Titanium:			
		hkl,	E,	ν
		100,	110.86,	0.3290
		002,	128.35,	0.2976
		101,	123.41,	0.3061
		102,	126.83,	0.3002
		Macro,	114.70,	0.3217

In this measurement, E and ν were constant for temperature, and thermal expansion coefficient α was considered for the temperature dependency. Specifically, these temperature dependencies were defined by results of the experimental measurement for titanium and tungsten in this study. The measurement time for one diffraction profile was about 10 min for each of the three samples explained in Figure 5. Lastly, the measuring time of one cycle in this measurement became about 4 h.

Furthermore, the simple elastic calculations of thermal expansions were performed and compared with experimental results. This calculation assumes the one axial loading model, and it is shown by the following theoretical equation:

$$\begin{aligned} \sigma_W &= \frac{E_W E_{Ti} V_{Ti}}{E_W V_W + E_{Ti} V_{Ti}}(\alpha_W - \alpha_{Ti})\Delta T \\ \sigma_{Ti} &= \frac{E_W E_{Ti} V_W}{E_W V_W + E_{Ti} V_{Ti}}(\alpha_{Ti} - \alpha_W)\Delta T, \end{aligned} \quad (2)$$

where suffixes denote the materials, E is Young's modulus, α is the coefficient of thermal expansion, ΔT is the temperature difference, and V is the volume fraction of the tungsten fiber. Young's modulus E and the thermal expansion coefficient α used the same parameters in Table 2, and the volume fraction was V_W = 5% in this calculation. The initial stresses needed to calculate the residual stresses were supposed at the room temperature 300 K position. The measurement values in both the tungsten fiber and the titanium matrix were used for these initial residual stresses.

3. Results

3.1. Diffraction Profile by the Time-of-Flight Method

Figure 6 shows an example of the diffraction profile measured by the TOF method of TAKUMI. In this profile data, diffraction peaks from the tungsten fibers and the titanium matrix of the W/Ti composite appeared, and diffraction peaks from the copper plate on which the samples were attached could be confirmed. This diffraction profile was obtained in a measurement time of 10 min. A very clear and wide range peak profile was obtained. This is a measurement result that cannot be obtained using the angular dispersion method with a goniometer, which the author has previously used. These diffraction profiles were peak-fitted using the Z-Rietveld software for Rietveld analysis [18]. However, when there are peak shifts in profiles, some peaks appear at irregular positions, and Z-Rietveld software can be fitted for any peaks independently of the constraints from Bragg's equation. Therefore, the Z-Rietveld software was employed for the peak-fitting treatment in this study. The Z-Rietveld software is provided free of charge from the JAEA homepage.

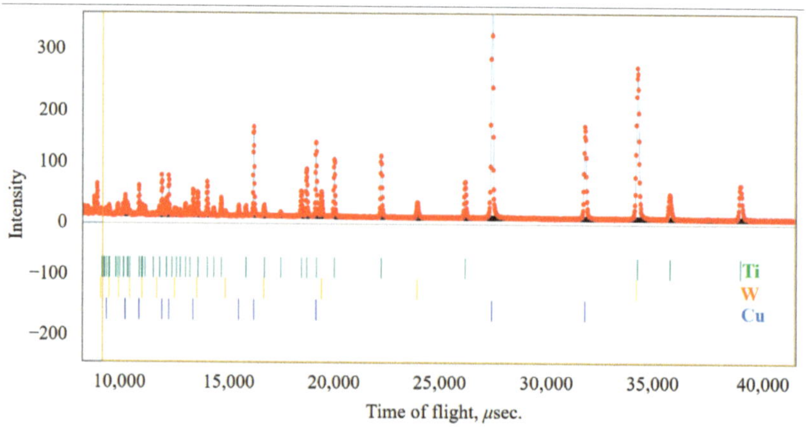

Figure 6. One example of the diffraction profile measured using the time-of-flight method of TKUMI. These diffraction profiles were peak-fitted using the Z-Rietveld software.

Figure 7a,b show the comparison of the diffraction profile in the N-direction and the S-direction from W/Ti composites. In this figure, diffraction peaks from W110, W200, W220,

Ti101, and Ti202 are compared. Figure 7a is the result of measurement in the N-direction, where W110 and W220 appear and W200 disappears. On the other hand, Figure 7b is the result of measurement in the S-direction, where W110 and W220 disappear and W200 appears. From these results, it was confirmed that the tungsten fibers have a strong 110 fiber preferential orientation form these results.

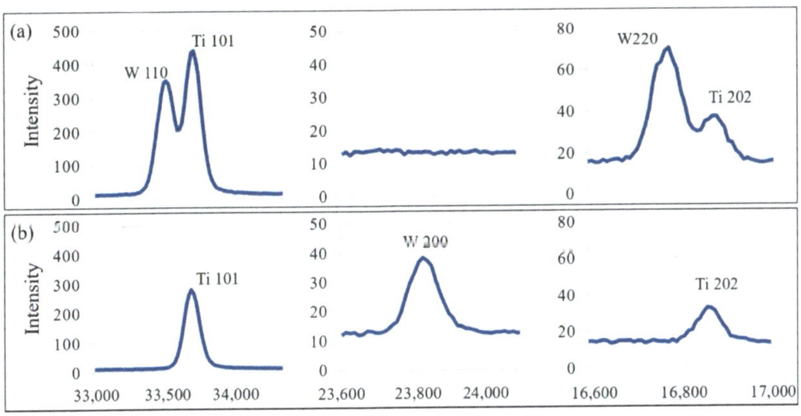

Figure 7. Comparison of the diffraction peaks in N-direction and S-direction. (**a**) N-direction: W110 and W220 appear. (**b**) S-direction: W110 and W220 disappear, and W200 appears.

Previous reports also confirmed the existence of a strong 110 fiber preferential orientation in tungsten fibers. Furthermore, the 110 fiber preferential orientation of tungsten fibers was also confirmed in the raw material tungsten fibers before W/Ti production. The main cause is considered to be the drawing process when the fibers were manufactured. On the other hand, such an orientation was not confirmed in the diffraction line of titanium matrix. In this way, the disappearance of the diffraction peak due to the fiber preferential orientation is a serious problem when measuring with an angle dispersion goniometer. Due to the extremely long measurement time required by the angular dispersion method, it was not possible to measure the entire range of diffraction peaks in advance. However, the TOF method of TAKUMI can simultaneously measure many diffraction peaks in addition to the disappeared peak W200. Although some diffraction peaks of tungsten fibers disappeared in this W/Ti composite, tungsten has the properties of being completely isotropic, and it was assumed that the lattice spacing d measured at any diffraction plane would be the same value. Therefore, the average strains calculated from all tungsten diffraction peaks fitted by the Z-Rietveld software were evaluated in this study.

3.2. Strain Calculation by the TOF Method

In this study, the strains generated by temperature alterations were calculated from the measurement results of the TOF method of TAKUMI. The relationship between lattice spacing d and flight time t in the time-of-flight method is explained by the following equation

$$d = \frac{\lambda}{2sin\theta} = \frac{1}{2sin\theta}\frac{h}{mL}, \quad (3)$$

In this equation, h is Planck's constant, m is the mass of a neutron, and L is the flight distance of neutrons, which is a constant value at the distance from the neutron source to TAKUMI (BL19). The diffraction angle θ is also a constant specific to the TAKUMI instrument. From Equation (3), it can be confirmed that the lattice spacing d is simply related to the flight time t by a constant value.

$$\varepsilon = \frac{d - d_0}{d_0} = \frac{TOF - TOF_0}{TOF_0}. \quad (4)$$

The strain ε in each direction is given by Equation (4). By substituting Equation (3) into Equation (4), the strain can be calculated using the flight time t. Since there is no need to convert to the lattice spacing d, the strain can be calculated very easily. From this result, strains were calculated directly from the measured values of the time of flight t in this study. The strain measurements are basically the relative changes of peak positions from the stress-free condition; therefore. as long as the data reduction processes are conducted with the same procedure for all conditions, the effect of data reduction in the peak position must be very small [19]. These experiments focused on the thermal stresses, and the change in the peak symmetry by cooling and heating was not observed. The change in the position of the measurement was performed only by changing the sample types (composite and two constituents as d_0 samples) when the temperature was changed. For the same sample, the measurement was performed exactly at the same position. Moreover, the data analysis was conducted to refine the lattice constant using the Pawley method with many peaks; therefore. the effect of the energy distribution was also very small.

3.3. Results of Thermal Strains

Figure 8a,b show examples of thermal strain alterations from the results of the diffraction peaks measurement of the titanium matrix and the tungsten fibers in the W/Ti composite. These thermal strains were caused by thermal shrinkage in the longitudinal direction of fibers. These strains were calculated using the lattice spacing at room temperature as the initial value. However, the lattice spacing was not actually obtained; the TOF value was used as described above.

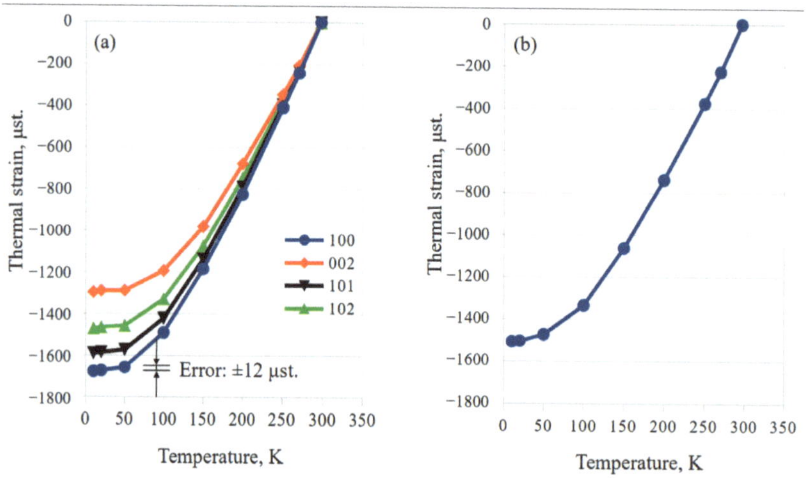

Figure 8. Alterations of thermal strains determined from the results of the diffraction peak measurements of (**a**) the titanium matrix and (**b**) the tungsten fibers in the W/Ti composite.

From this result, it can be confirmed that thermal strains were generated from thermal shrinkage due to temperature changes in both the titanium matrix and the tungsten fiber. Figure 8a shows the measurement results for titanium matrix. From these results, it can be confirmed that different thermal strains were generated in the titanium matrix depending on the *hkl* diffraction planes. Since the crystal system of α-titanium is a hexagonal structure, the thermal expansion coefficients of the a-axis and the c-axis are different. Therefore, different thermal strains were generated in each diffraction plane. In Figure 8a, the 100 plane with the largest strain corresponds to the a-axis of thermal strains, and the 002 plane with the smallest value corresponds to the c-axis in thermal strains. As the hexagonal crystal has many diffraction planes, thermal strains from other diffraction planes exist between the thermal strains of the a-axis and the c-axis. The Ti101 plane and the Ti102 planes are shown in Figure 8a as representatives. Error bars are displayed for the measurement results of the Ti100 plane in Figure 8a. Error bars indicate the scatter of thermal strain at each temperature. It can be inferred that the measurement was performed with high precision because the scatter of strains in each temperature was extremely small about ± 12 μ strain. From the results of this measurement, it was confirmed that the alterations in thermal strain were generated by the thermal cycle changes along the same path in the case of temperature rise and temperature drop. These tendencies were the same for other diffraction planes, and the scatter in thermal strain at each temperature was very small. Therefore, the display of error bars on other diffraction planes was omitted.

Figure 8b shows the thermal strain alterations in the tungsten fibers. Only the average value is shown because the scatter in the strain value at each temperature was even smaller than the measurement results for titanium matrix. In the case of tungsten, since there is no anisotropy due to the diffraction planes, the thermal strains do not depend on the diffraction planes, and the same result was obtained regardless of which diffraction planes were selected. From both results, it was confirmed that the alterations in thermal strain were generated by the thermal cycle changes along the same path in the case of temperature rise and temperature drop. These results show the strain in both the titanium matrix and the tungsten fibers fluctuated the region of elastic deformation, indicating that plastic deformation does not occur.

3.4. Results of Thermal Stress Alterations

Figure 9a, b show the stress alterations of the titanium matrix in the W/Ti composite. Figure 9a is the result of the longitudinal direction of tungsten fibers, and Figure 9b is the normal direction. Both of these figures show the stress values calculated from the 100 plane (a-axis), the 002 plane (c-axis), and the average stress values obtained from the four diffraction planes. According to Figure 9a in the longitudinal fiber direction, the 100 plane has the lowest stress values and the 002 plane has the highest stress values. The initial residual stress is 33 MPa on the 100 plane and 82 MPa on the 002 plane. The average value of the four diffraction planes is 57 MPa, which is the intermediate value of the initial residual stress of the 100 plane and the 002 plane. It can be inferred that these initial residual stresses were the thermal residual stresses generated in the process of manufacturing the W/Ti composite. Such initial residual stresses always occur in composite materials.

Thermal residual stresses shift to tensile stresses with decreasing temperature. The tendency of increase was qualitatively the same for all diffraction planes. The alteration of the stress value obtained from the average stress was 12 MPa during the temperature change from room temperature to 10 K. According to Figure 9b, all stress values in the normal to the fiber direction were compressive residual stresses. The tendency of stress alterations was the same as Figure 9a; it shifted to the tensile stress side and compressive stress decreased with decreasing temperature. The stress alteration from room temperature to 10 K was also the same value of 12 MPa.

From these results, it was confirmed that the stress alterations on the order of 10 MPa could be evaluated with high accuracy. This high measurement accuracy is an advantage of the TOF method in TAKUMI. This is an order of accuracy which cannot be obtained with the conventional angular dispersion type method using a goniometer.

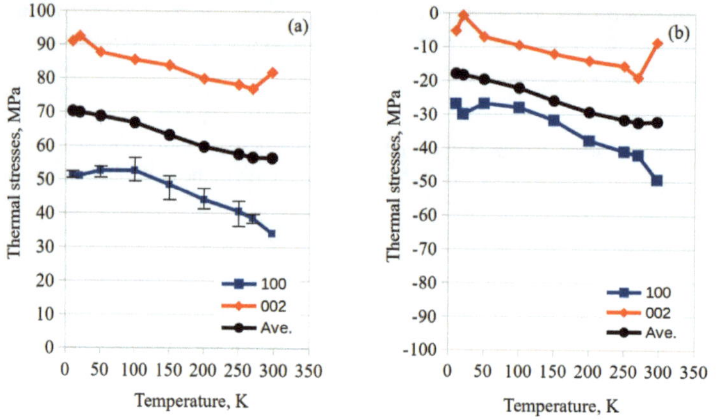

Figure 9. Stress alterations of the titanium matrix in the W/Ti composite: (**a**) the longitudinal direction of tungsten fibers; (**b**) the normal direction.

Figure 10a,b show the stress alterations of tungsten fibers in W/Ti composites. Figure 10a is the longitudinal direction of the tungsten fiber, and Figure 10b is the normal direction. In the case of tungsten fiber, the intensity of the diffraction peak in the normal direction is very weak because of the 110 fiber preferential orientation. Therefore, the measurement results of each temperature show some scatter indicated by the error bars in this figure. According to Figure 10a, the thermal residual stresses in the fiber longitudinal direction were all compressive states. The initial residual stress was −963 MPa, and thermal residual stresses shifted to compressive stress with decreasing temperature. The alteration in the stress value was about 110 MPa between room temperature and 10 K temperature change. It was confirmed that the stress alteration in the tungsten fiber was approximately 10 times compared with the stress alteration of the titanium matrix. On the other hand, according to Figure 10b, all stress values in the normal to the fiber direction were tensile residual stress states. The initial residual stress was about 50 MPa. Although it is difficult to confirm from this figure, the stress alterations accompanying the decrease in temperature shifted slightly to the tensile side. The alteration in stress value was about 10 MPa.

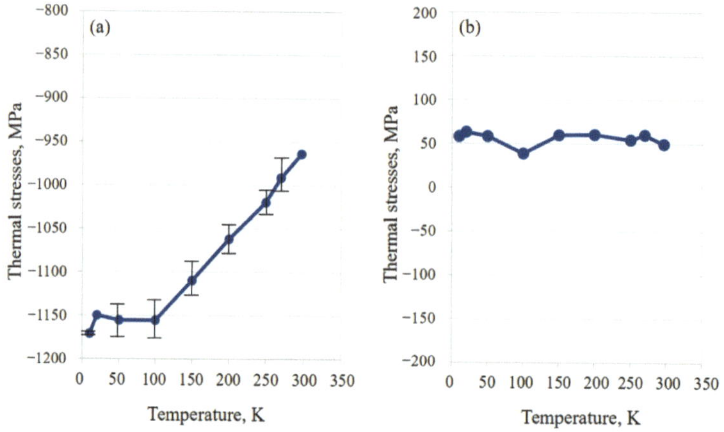

Figure 10. Stress alterations of tungsten in W/Ti composites: (**a**) the longitudinal direction of the tungsten fiber; (**b**) the normal direction.

From the results in Figures 9a and 10a, comparing the stress states of the titanium matrix and the tungsten fibers in the longitudinal fiber direction, it can be found that the titanium matrix had tensile stresses and the tungsten fiber had compressive stresses. In addition, comparing the magnitude of stresses between the longitudinal fiber direction and the normal direction, it can be confirmed that the stress alterations in the longitudinal fiber direction was very large and dominant. It can be inferred that such states of thermal residual stresses were caused by the mismatch of thermal expansion between the titanium matrix and the tungsten fiber.

In the fiber longitudinal direction, since the thermal expansion of the titanium matrix was larger than that of the tungsten fiber, tensile residual stress was generated in the titanium matrix and compressive residual stress was generated in the tungsten fiber when the temperature decreased.

Conversely, in the fiber normal direction, the titanium matrix was in compression, and the tungsten fiber was in tension. These phenomena were opposite for the difference in thermal expansion. Since the stress value was small, it can be viewed as a measurement error. However, since it is not possible to examine this in detail at this time, it will be left as a future study.

4. Discussion

Figure 11a,b show a comparison between the results of thermal stresses calculated by Equation (2) and the measured results using the TOF method of TAKUMI. These results are in the longitudinal fiber direction. In the thermal stress calculations, the thermal expansion coefficients of the titanium matrix and tungsten fiber were calculated from the results of thermal strains measured from the d0 sample. For the Young's modulus E, the macro of Young's modulus in Table 2 was used for the thermal stress calculations for the titanium matrix. The initial residual stresses of the titanium matrix and tungsten fibers were $\sigma_{\text{Ti-initila}}$ = 56 MPa for titanium and $\sigma_{\text{W-initial}}$ = −963 MPa for tungsten from the result of the stress measurement.

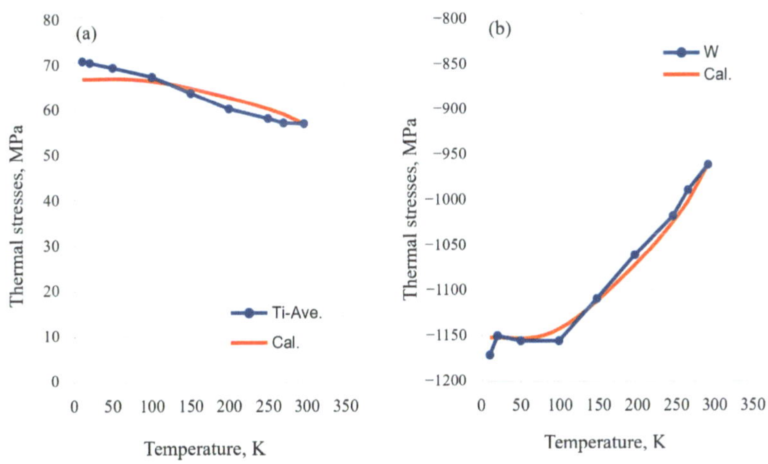

Figure 11. Comparison of calculated results from Equation (2) and measurement results in the longitudinal fiber direction: (**a**) results of the titanium matrix; (**b**) results of the tungsten fibers.

Figure 11a compares the calculated and measured thermal stresses in the titanium matrix. The average stresses of the six diffraction planes are used as a representative for the actual measurement results by the TOF method. First, the validity of the absolute values of the initial stress $\sigma_{\text{Ti-initial}}$= 56 MPa of the titanium matrix and the initial stress $\sigma_{\text{W-initial}}$= −963 MPa of the tungsten fiber obtained from the measurement results is discussed. In general, residual stress in composite materials is considered to be dynamically balanced inside the material. The internal stress is balanced by being distributed according to the volume fraction of each phase composing the material. In the case of W/Ti composites, it is believed that the residual stresses of the titanium matrix and tungsten fibers are balanced according to the volume fraction of the titanium matrix and the tungsten fibers. This compound rule is represented by the following formula:

$$\sigma_{Ti} \cdot V_{Ti} + \sigma_W \cdot V_W = 0 \ , \tag{5}$$

where V is the volume fraction for the titanium matrix and tungsten fibers in the W/Ti composite, and the suffix indicates each material. Since the volume fraction of tungsten fibers in the W/Ti composite in this study was about 5%, the first term of the Equation (5) is $\sigma_{Ti} \times V_{Ti} = 56 \times 0.95 = 53.2$ and the second term is $\sigma_W \times V_W = 963 \times 0.05 = 48.15$, which are very close values. According to the calculation result of Equation (5), the calculated volume fraction of tungsten fibers is slightly larger than 5%, which was 5.5%. For example, if the volume fraction of tungsten fibers is increased or decreased by 1%, the above magnitude relationship of stress values is a large variation. From this result, it can be confirmed that these measurements succeeded in accurately evaluating the stress balance between the titanium matrix and the tungsten fibers in this measurement.

Next, a comparison between the calculated thermal stresses and the measured values is considered. According to Figure 11a, when comparing the calculated and measured thermal expansion results, thermal residual stresses shifted to tensile stress when the temperature decreased in both calculated results and measured results. Although the tendencies of the experimental results and the calculated results coincided qualitatively, the absolute values of the calculated results and the measured results did not align at the same at each temperature when strictly compared. However, the difference between the calculation result and the actual measurement result at the lowest temperature of 10 K was only about 4 MPa. A higher accuracy is difficult to obtain in neutron stress measurements.

As shown in Figure 11b, the calculated and measured stress alterations in the tungsten fiber were in very good agreement. Since the scale on the vertical axis in Figure 11b is five times that of the titanium matrix in Figure 11a, it is difficult to read the difference in absolute values from this figure. However, the maximum difference between calculated and measured results was 22.8 MPa at 10 K. This value can also be said to be a very small difference between the measured results and the calculated results. The value of residual stress measured and/or calculated in this study must be added to the stress value that can be applied in strength design. In particular, it is important to consider the titanium matrix side where tensile residual stress is generated by cooling.

From the above results, it was confirmed that the stresses in the fiber longitudinal direction of the W/Ti composite coincided very well with the calculated results in terms of stress alterations in both the titanium matrix and the tungsten fibers. These calculations were obtained from a simple elastic theory of thermal expansion of titanium matrices and tungsten fibers in Equation (2). It was also confirmed that the measurement accuracy of the time-of-flight method using TAKUMI is extremely high.

5. Conclusions

(1) Regarding the initial residual stress in the fiber longitudinal direction, compressive stresses existed in tungsten fibers, and tensile stresses existed in the titanium matrix.
(2) Thermal residual stresses in tungsten fibers and the titanium matrix were changed to other states depending on temperature changes.
(3) The main factor of thermal stress alterations was the difference in thermal expansion between tungsten fiber and titanium matrix, and the effect in the longitudinal direction of the fibers was dominant.
(4) The alterations in thermal stresses followed the same path during temperature rise and temperature drop in the thermal cycle changes.
(5) The simple elastic calculations and the measured results showed very good agreement, confirming the high measurement accuracy of the time-of-flight method using TAKUMI.

Author Contributions: S.H., T.K., T.Y. and W.G. are JAEA staff members. They set up the samples for TAKUMI and operated all equipment. M.N. prepared the samples, analyzed the experimental results, and wrote the manuscript. All authors discussed the results and reviewed the manuscript. All authors have read and agreed to the published version of the manuscript.

Funding: This research received no external funding.

Data Availability Statement: The data presented in this study are available on request from the corresponding author.

Acknowledgments: Neutron experiments were performed at TAKUMI installed at beamline BL19 in J-PARC with the approval of the Japan Atomic Energy Agency (JAEA). Acceptance number: 2019B129.

Conflicts of Interest: The authors declare no conflict of interest.

References

1. Ogasawara, T.; Arai, N.; Fukumoto, R.; Ogawa, T.; Yokozeki, T.; Yoshimura, A. Titanium alloy foil-inserted carbon fiber/epoxy composites for cryogenic propellant tank application. *Adv. Compos. Mater.* **2013**, *23*, 129–149. [CrossRef]
2. El Etri, H.; Korkmaz, M.E.; Gupta, M.K.; Gunay, M.; Xu, J. A state-of-the-art review on mechanical characteristics of different fiber metal laminates for aerospace and structural applications. *Int. J. Adv. Manuf. Technol.* **2022**, *123*, 2965–2991. [CrossRef]
3. Smit, T.C.; Nobre, J.P.; Reid, R.G.; Wu, T.; Niendorf, T.; Marais, D.; Venter, A.M. Assessment and Validation of Incremental Hole-Drilling Calculation Methods for Residual Stress Determination in Fiber-Metal Laminates. *Exp. Mech.* **2022**, *62*, 1289–1304. [CrossRef]
4. Ramesh, K.C.; Sagar, R. Fabrication of Metal Matrix Composite Automotive Parts. *Int. J. Adv. Manuf. Technol.* **1999**, *15*, 114–118. [CrossRef]
5. Okamoto, S.; Taki, M.; Hagihara, T.; Fuyama, N. Material Properties on Metal Matrix Composite Used as Reinforcement for Pistons of Engines in Automobiles. *Trans. Jpn. Soc. Mech. Eng. Ser. A* **2007**, *73*, 368–371. [CrossRef]
6. Asanuma, H.; Kato, K.; Mochizakl, T.; Hakoda, G.; Kurihira, H. Development of active materials based on fiber-reinforced metals. *J. Adv. Sci.* **2006**, *18*, 16–19. [CrossRef]
7. Emamian, S.S.; Awang, M.; Yusof, F.; Sheikholeslam, M.; Mehrpouya, M. Improving the friction stir welding tool life for joining the metal matrix composites. *Int. J. Adv. Manuf. Technol.* **2020**, *106*, 3217–3227. [CrossRef]
8. Hua, M.; Guo, W.; Law, H.W.; Ho, J.K.L. Half-transient liquid phase diffusion welding: An approach for diffusion welding of SiCp/A356 with Cu interlayer. *Int. J. Adv. Manuf. Technol.* **2007**, *37*, 504–512. [CrossRef]
9. Ekin, J.W. *Experimental Techniques for Low Temperature Measurements*; Oxford University Press: Oxford, UK, 2008; pp. 50–52, 234–235.
10. Network Database System for Thermophysical Property Data, The National Institute of Advanced Industrial Science and Technology (AIST). Available online: https://tpds.db.aist.go.jp/index_en.html (accessed on 5 January 2023).
11. Nishida, M.; Haneoka, M.; Matsue, T.; Jing, T.; Hanabusa, T. Thermal Stress Estimation of Tungsten Fiber Reinforced Titanium Composite by In-situ X-ray Diffraction Method. *Mat. Sci. Forum* **2014**, *768–769*, 335–342.
12. Ikeuchi, Y.; Matsue, T.; Hanabusa, T. X-ray Triaxial Evaluation of Thermal Residual Stresses in Continuous Alumina Fiber Reinforced Aluminum. In Proceedings of the 5th International Conference on Residual Stresses, Linköping, Sweden, 16–18 June 1997; Volume 2, pp. 958–963.
13. Ikeuchi, Y.; Iga, K.; Hanabusa, T. X-ray Triaxial Evaluation of Thermal Residual Stress in Continuous Alumina Fiber Reinforced Aluminum Composite. *J. Soc. Mat. Sci. Jpn.* **1994**, *43*, 792–798. (In Japanese) [CrossRef]
14. Nishida, M.; Muslih, M.R.; Ikeuchi, Y.; Minakawa, N.; Hanabusa, T. Low temperature in-situ stress measurement of w/cu composite by neutron diffraction. *Int. J. Mod. Phys. B* **2006**, *20*, 3668–3673. [CrossRef]
15. Masayuki, N.; Muslih, M.; Ikeuchi, Y.; Minakawa, N.; Takao, H. Internal Stress Measurement of Fiber Reinforced Composite by Neutron Diffraction with In Situ Low Temperature Stress Measurement System. *Mater. Sci. Forum* **2005**, *490–491*, 239–244. [CrossRef]
16. Clausen, B.; Brown, D.W.; Noyan, I.C. Engineering Applications of Time-of-Flight Neutron Diffraction. *JOM* **2012**, *64*, 117–126. [CrossRef]
17. X-ray Elastic Constants by Krouner Model by The Committee on X-ray Study on Mechanical Behavior of Materials, Japan. Available online: https://x-ray.jsms.jp/x_database/index.html (accessed on 5 January 2023).
18. Oishi, R.; Yonemura, M.; Nishimaki, Y.; Torii, S.; Hoshikawa, A.; Ishigaki, T.; Morishima, T.; Mori, K.; Kamiyama, T. Rietveld analysis software for J-PARC. *Nucl. Instrum. Methods Phys. Res. Sect. A: Accel. Spectrometers Detect. Assoc. Equip.* **2008**, *600*, 94–96. [CrossRef]
19. Harjo, S.; Moriai, A.; Torii, S.; Suzuki, H.; Suzuya, K.; Morii, Y.; Arai, M.; Tomota, Y.; Akita, K.; Akiniwa, Y. Design of Engineering Diffractometer at J-PARC. *Materi. Sci. Forum* **2006**, *524*, 199–204. [CrossRef]

quantum beam science

Article

Stress Evaluation Method by Neutron Diffraction for HCP-Structured Magnesium Alloy

Stefanus Harjo *, Wu Gong and Takuro Kawasaki

J-PARC Center, Japan Atomic Energy Agency, Tokai-mura, Naka-gun, Ibaraki 319-1195, Japan
* Correspondence: stefanus.harjo@j-parc.jp

Abstract: Tensile deformation in situ neutron diffraction of an extruded AZ31 alloy was performed to validate conventional procedures and to develop new procedures for stress evaluation from lattice strains by diffraction measurements of HCP-structured magnesium alloys. Increases in the lattice strains with respect to the applied true stress after yielding largely vary among [*hk.l*] grains. Some [*hk.l*] grains have little or no increase in lattice strain, making it difficult to use the conventional procedures to determine the average phase strain by using lattice constants or by averaging several lattice strains. The newly proposed procedure of stress evaluation from the lattice strains shows very high accuracy and reliability by weighting the volume fraction of [*hk.l*] grains and evaluating them in many [*hk.l*] orientations in addition to multiplication by the diffraction elastic constant. When multiple *hk.l* peaks cannot be obtained simultaneously, we recommend to use the 12.1 peak for stress evaluation. The lattice strain value evaluated from the 12.1 peak shows a good linear relationship with the applied true stress for the whole deformation region.

Keywords: magnesium alloy; neutron diffraction; stress evaluation; tension; in situ

1. Introduction

Magnesium (Mg) and its alloys are the lightest structural metallic materials and have a great potential in various applications. They are already used for automotive applications [1,2] and are also candidate materials for aircraft applications. Residual stress measurements on the engineering parts made of Mg alloys are, therefore, very important to guarantee quality and safety. Recently, Mg alloys have become multi-phase [3–5] in order to achieve higher performances. For further developments, it has become necessary to understand the deformation behavior of each of the constituent phases, including phase stresses and intergranular stresses. One of the powerful probes to evaluate the residual stresses and the stresses in the constituent phases during a process is the diffraction method using a quantum beam (X-ray or neutron).

X-ray diffraction with the $\sin^2\psi$ method [6] as well as the triaxial-stress method [6] are widely used to measure the residual stresses on the surface of engineering parts [7,8]. Neutron diffraction [9] is used to measure the residual stresses at the inside region of engineering parts [10] and also the evolutions of stresses in situ during deformation [11,12]. Diffraction occurs on multiple (*hkl*) lattice planes within the crystal, and the stress measurement is performed based on a change in the spacing of the (*hkl*) lattice plane from the stress-free state. The evaluated stress value varies depending on the *hkl* peak or (*hkl*) lattice plane [13], particularly in the engineering parts received plastic forming or in the samples deformed in the plastic region. Crystal grains have different deformabilities depending on their [*hkl*] orientations and share stresses differently after yielding. The selection of the *hkl* peak is, therefore, very important to estimate the stress value, representing the macroscopic one or the phase average one.

The structural metallic materials share stresses among the [*hkl*] orientations during deformation due to the elastic–plastic anisotropy of the [*hkl*] orientation. For materials having FCC or BCC structures, the change in [*hkl*] stress in response to external forces remains of the same sign even when stress sharing occurs among the [*hkl*] orientations [11,14]. In the FCC- or BCC-structured material, the 311 or 211 peak, respectively, is often used for the stress evaluation [13]. When multiple peaks are measured simultaneously using neutron diffraction with the time-of-flight method, the average lattice constants refined from the analyses using the Rietveld or Pawley methods [15,16] are also used for the stress evaluation. For Mg alloys having an HCP structure, the changes in the [*hk.l*] stress with respect to the external forces vary drastically when yielding occurs; some of their slopes changed to the opposite signs [12,17–19]. Therefore, we need to clarify whether the similar methods for FCC- and BCC-structured materials can be used or not. The lattice constants a and c, refined using the Rietveld method [20,21], and the spacing of the 10.1 or 10.3 lattice plane [20,22] are often used for the stress evaluation. In the X-ray diffraction $\sin^2\psi$ method, the 12.3 peak has been recommended for stress measurement in HCP-structured materials [23] for over three decades. The reason for this 12.3 peak selection is, however, unclear. Moreover, the validation of the measured stress values by diffraction for HCP-structured materials has not been performed well.

In this study, hence, we perform tensile deformation in situ neutron diffraction of an AZ31 alloy, monitor the crystallographic evolutions as deformation progresses, evaluate the stress values according to the conventional procedures described above, and compare the evaluated stress values with the macroscopic stress values obtained mechanically. We also propose new procedures to evaluate the stress value that represents macroscopically or as the phase average value.

2. Materials and Methods

The sample used in this study is a commercial AZ31 alloy, which was extruded with the extrusion ratio of 10 at 623 K. A dog bone specimen with the active length and diameter of 15 mm and 4 mm, respectively, was prepared in such a way that the loading direction was parallel to the extrusion direction.

An in situ neutron diffraction experiment during a uniaxial tensile test at room temperature was performed using TAKUMI [24], the neutron diffractometer with the time-of-flight method for engineering materials at Materials and Life Science Experimental Facility [25] of Japan Proton Accelerator Research Complex. The loading axis was aligned 45° horizontal to the incident beam. The details on the alignment can be found elsewhere [11,14,19,26,27]. Using a pair of 90° scattering detector banks in TAKUMI, the diffraction pattern with the scattering vector parallel to the loading axis that, hereafter, is referred to as the loading direction (LD), and perpendicular to the loading axis that, hereafter, is referred to as the transverse direction (TD) was simultaneously collected. Figure 1a shows the geometric schematic view of the in situ neutron diffraction experiment during the tensile test.

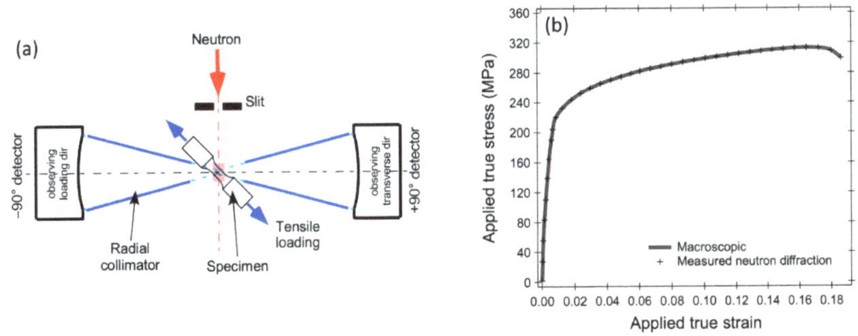

Figure 1. (**a**) Geometric schematic view of the in situ neutron diffraction measurement during the tensile test. (**b**) Applied true stress–true strain curve of extruded AZ31 with the tensile test.

The tensile test was performed as follows. In the elastic regime, the loads were applied in stages, and in the plastic regime, deformation was applied continuously at a constant displacement rate (strain rate of approximately 1×10^{-5} s^{-1}). The neutron diffraction data were collected continuously during the tensile test by using an event data recording mode [28,29]. The strain was monitored with a digital image correlative method (DIC) on images of the specimens captured every 5 s [30]. The tensile test was performed until the specimen fractured. The applied true stress–true strain curve is shown in Figure 1b. Localized strains were not observed during the test up to the ultimate tensile strength, and the applied true stresses and true strains were calculated from the applied engineering stresses and engineering strains using the uniform deformation assumption. The neutron diffraction data were extracted corresponding to the load holding for the elastic regime and periodically at time intervals of 300 to 600 s for the plastic regime. A single peak fitting procedure to obtain peak integrated intensities and peak positions was performed using the Z-Rietveld software, v. 1.1.10 [31]. A Rietveld texture analysis [32,33] to evaluate the texture condition was also performed using the MAUD software, v. 2.78 [34].

3. Results and Discussion

3.1. Responses of [hk.l] Orientations

Figure 2a shows the diffraction patterns before deformation in the LD and TD. A crystal model for Mg with the space group of $P6_3/mmc$ and the lattice parameters of a = 0.320 nm and c = 0.518 nm was used to simulate the peak positions, which are shown as tick marks in Figure 2a. All peak positions are in good agreement with the tick marks, showing that the specimen is a single-phase, HCP-structured Mg alloy. The peak intensity ratio in the LD is different from that in the TD, exhibiting the presence of texture. Figure 2b shows the inverse pole figures (IPF) in the LD before deformation. The IPF were obtained from the Rietveld-texture analysis [32,33], assuming a fiber texture, using the twelfth-order spherical harmonic function implemented in the MAUD software, v. 2.78 [34]. The values of multiple random distributions (mrd) are large near the [10.0] pole, displaying that a typical basal texture with the c-axis orthogonal to the extrusion direction [12,19,26,35,36] was developed. The condition of the IPF, in which the mrd values are large near the [10.0] pole, is kept even at the applied true strain value of 16.4%, i.e., the change in texture is small.

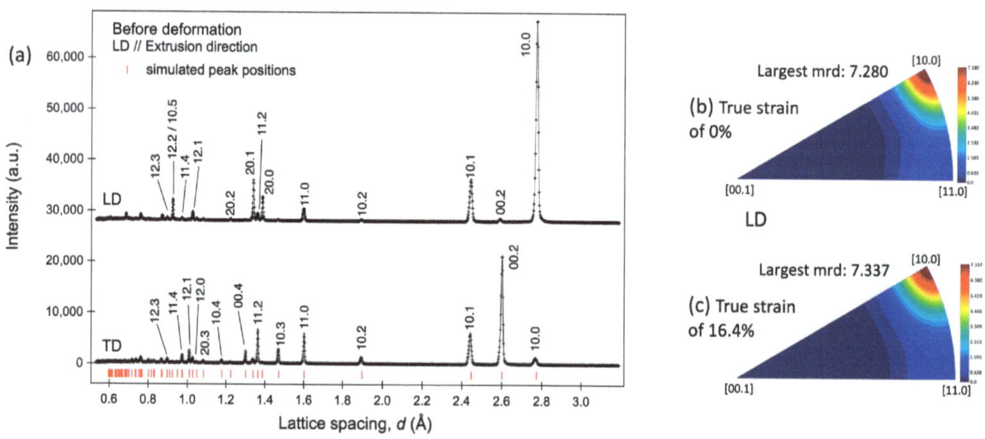

Figure 2. (**a**) Diffraction patterns of extruded AZ31 before deformation in the loading direction and the transverse direction. The extrusion direction is parallel to the loading direction. (**b**) Inverse pole figure in the loading direction before deformation and (**c**) that at the applied true strain value of 16.4%.

Figure 3a,a′,b,b′ show the relative integrated intensities of several *hk.l* peaks. The relative integrated peak intensity was estimated by normalizing the *hk.l* integrated peak intensity measured during deformation to the *hk.l* integrated peak intensity before deformation, and it can be used to understand the evolution of texture by deformation. As can be observed in Figure 3a, in the LD, the relative integrated intensities of the 00.2 and 10.3 peaks decrease with the increase in the applied true stress above 210 MPa, accompanied by a slight increase in the integrated peak intensity of 10.0, and vice versa in the TD shown in Figure 3b. These integrated peak intensity changes with respect to the applied true stress may display the occurrence of twinning [37,38]. The integrated intensities of the 00.2 and 10.3 peaks in the LD before deformation are very small, as shown in Figure 2a; therefore, the relative integrated peak intensity changes of the 00.2 and 10.3 peaks in the LD are very sensitive.

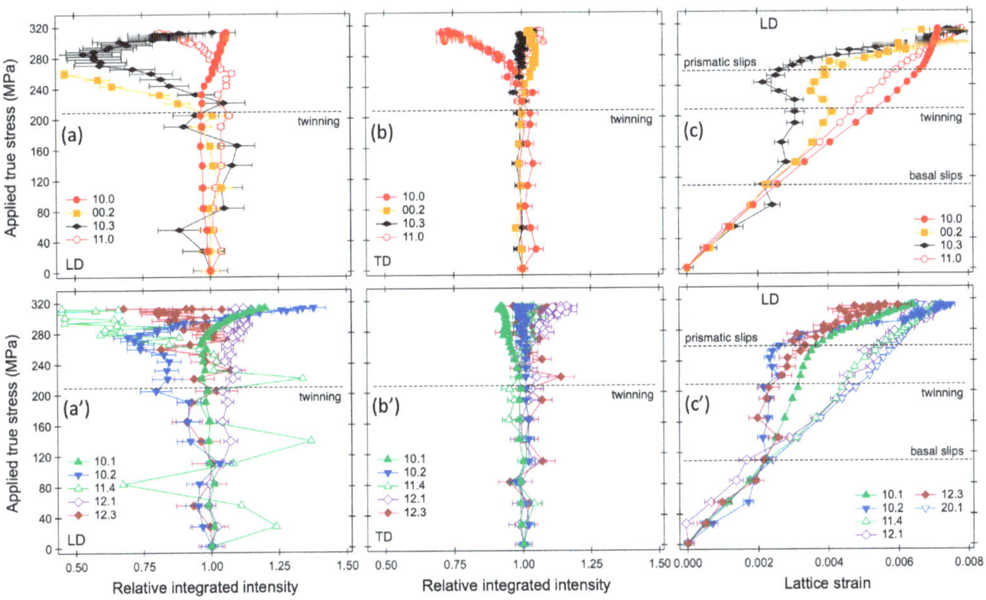

Figure 3. Relative integrated intensities of several *hk.l* peaks in (**a**,**a′**) the loading direction and (**b**,**b′**) the transverse direction. (**c**,**c′**) Lattice strains of several [*hk.l*] oriented grains in the loading direction.

Figure 3c,c′ show the lattice strains of several [*hk.l*] oriented grains ($\varepsilon^{hk.l}$) in the LD. The lattice strain was estimated according to the following equation:

$$\varepsilon^{hk.l} = \left(d^{hk.l} - d_0^{hk.l}\right) \Big/ d_0^{hk.l}, \quad (1)$$

where $d^{hk.l}$ and $d_0^{hk.l}$ are the lattice spacings for [*hk.l*] oriented grains measured during deformation and before deformation, respectively. The occurrences of basal slips, twinning, and prismatic slips can be judged from the changes in the $\varepsilon^{hk.l}$ values with respect to the applied true stress. When basal slip occurs, the changes in the $\varepsilon^{hk.l}$ values with respect to the applied true stress of 10.1 and 10.2 become smaller [17,27,39]. The details about deformation modes in Mg alloys using in situ neutron diffraction, however, will not be discussed here because similar studies with detailed discussions have been reported elsewhere [17,35,37,40]. It is worth noting here that the linear relations between the applied true stress values and the $\varepsilon^{hk.l}$ value are difficult to be found in [10.0], [10.1], [11.0], and [20.1] oriented grains in the LD, having high peak intensities. The $\varepsilon^{hk.l}$ values for [12.3] oriented grains, which are recommended for the stress evaluation using X-ray diffraction [23], surprisingly show no linear relationship with the applied stress values throughout the deformation. The [12.3] oriented grains in the LD appear to yield due to basal slip and then bear stresses lower than the applied stress. Note also that the increases in the $\varepsilon^{hk.l}$ value with respect to the applied true stress for [10.2], [10.3], and [12.3] grains become very small after the initiation of basal slip; particularly, [10.2] grains have almost no increase in the slope.

3.2. *Conventional Procedures for Evaluations of Phase Strain and Phase Stress*

In many cases, average phase strains in HCP-structured materials have been conventionally calculated using the following two methods. One is a lattice strain analysis using the averaged lattice constants refined with the Pawley [41] or Rietveld method [42] as [21]:

$$\varepsilon^{\text{ave I}} = \varepsilon_{\text{HCP}}^{\text{lat.const.}} = \frac{2\left[\left(a^{\text{HCP}} - a_0^{\text{HCP}}\right) \Big/ a_0^{\text{HCP}}\right] + \left(c^{\text{HCP}} - c_0^{\text{HCP}}\right) \Big/ c_0^{\text{HCP}}}{3} = \frac{2\,\varepsilon_a + \varepsilon_c}{3}, \quad (2)$$

which is called the average I phase strain ($\varepsilon^{\text{ave I}}$). Here, a^{HCP} and c^{HCP} are the averaged lattice constants refined with the Pawley method using the Z-Rielveld software, v. 1.1.10 [31]. The other one is by simply averaging the $\varepsilon^{hk.l}$ values as [7,43]:

$$\varepsilon^{\text{ave II}} = \varepsilon_{\text{HCP}}^{\text{ave.}} = \frac{1}{n}\sum\nolimits_0^n \varepsilon^{hk.l}, \quad (3)$$

which is called the average II phase strain ($\varepsilon^{\text{ave II}}$).

Figure 4 shows the values of $\varepsilon^{\text{ave I}}$ and $\varepsilon^{\text{ave II}}$ in the LD, evaluated using the above two methods. For the comparison, the value evaluated from only the lattice parameter a (ε_a) is also plotted. None of them show a linear relation with the applied true stress for the whole deformation. Surprisingly, the value of $\varepsilon^{\text{ave I}}$, which is often used in the residual stress measurements with multiple *hkl* peaks [16,21,44,45], shows the largest deviation from the linear relation to have a smaller value, though the number of grains oriented to the LD where basal slips easily occur (e.g., [10.2] or [10.3] grains) is small due to the extrusion texture (see Figure 2a). The occurrence of basal slip still seems to have a strong influence on the average lattice constants. The linear response of $\varepsilon^{\text{ave II}}$ is kept up to the larger applied stress, but it corrupts when the applied stress value exceeds 160 MPa. The ε_a value is basically very similar to the $\varepsilon^{hk.l}$ value for [10.0] grains.

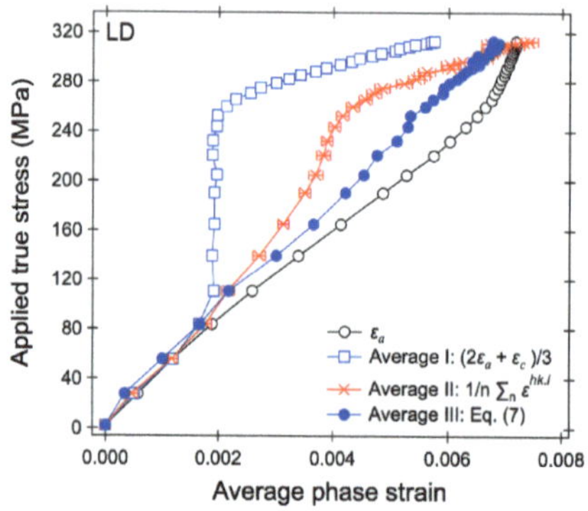

Figure 4. Average phase strain values evaluated using the conventional methods and the new proposed method of Equation (7).

Figure 5 shows the lattice stresses or phase stresses that were calculated conventionally from the $\varepsilon^{hk.l}$ values and the conventional average phase strain ($\varepsilon^{\text{ave I}}$, $\varepsilon^{\text{ave II}}$, or ε_a) values by multiplying them with the related diffraction elastic constants, according to the following equations:

$$\sigma^{hk.l} = \varepsilon^{hk.l}\ E^{hk.l}, \tag{4a}$$

$$\sigma^{\text{ave I}} = \varepsilon^{\text{ave I}}\ E^{\text{ave I}}, \tag{4b}$$

$$\sigma^{\text{ave II}} = \varepsilon^{\text{ave II}}\ E^{\text{ave II}}, \tag{4c}$$

$$\sigma_a = \varepsilon_a\ E_a. \tag{4d}$$

The value of the diffraction elastic constant was evaluated from the linear relation between the $\varepsilon^{hk.l}$, $\varepsilon^{\text{ave I}}$, $\varepsilon^{\text{ave II}}$, or ε_a and the applied true stress in the elastic regime. The evaluated values of $E^{hk.l}$ are shown in Figure 6a, and $E^{\text{ave I}} = \varepsilon^{\text{ave II}} = E_a = 45.2$ GPa. As shown in Figure 5a, the $\sigma^{hk.l}$ values evaluated for [*hk.l*] oriented grains, contributing to the large diffraction peak intensities in the LD, show good agreements only for the elastic region. The $\sigma^{hk.l}$ value for [10.1] grains deviates from the applied stress at a stress level much lower than the macroscopic yield stress due to the occurrence of basal slip. The $\sigma^{hk.l}$ value for [12.3] grains has been abbreviated here, but it is very close to the $\sigma^{hk.l}$ value for [10.1] grains, considering the results shown in Figure 3c. The other $\sigma^{hk.l}$ values for 10.0, 20.1, and 11.0 also receive influences differently when basal slips, twinning, or prismatic slips occur and deviate from the applied stress. The $\sigma^{\text{ave I}}$ and $\sigma^{\text{ave II}}$ values (Figure 5b) deviate also from the applied true stress at a very early deformation stage, which can be predicted also from Figure 4.

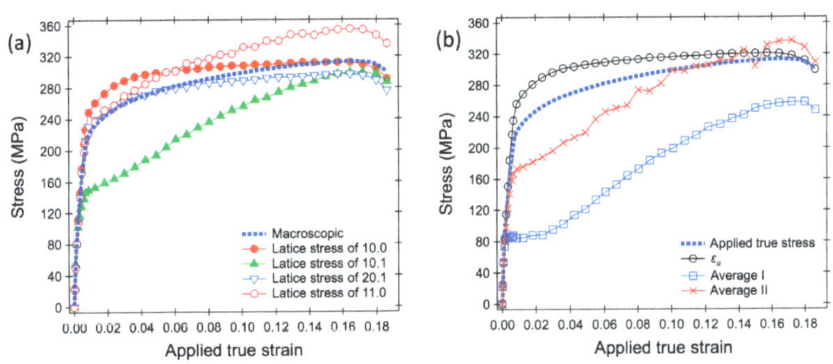

Figure 5. (**a**) Lattice stress values of several [*hk.l*] oriented grains and (**b**) average phase stress values estimated conventionally versus the applied true strains.

3.3. New Procedure for Phase Stress Evaluation

Here, we propose a new procedure to evaluate the stress value from the $\varepsilon^{hk.l}$ values by weighting the volume fraction of [*hk.l*] grains and evaluating them in many [*hk.l*] orientations in addition to multiplication by the diffraction elastic constants. The [*hk.l*] orientation with respect to the LD is replaced by an angle χ between the *c*-axis of the [*hk.l*] grains and the LD. The calculated stresses for the whole χ angles covering from 0° to 90° are then integrated. The equation for the phase stress evaluation is as follows:

$$\sigma = \int_{\chi=0^\circ}^{\chi=90^\circ} \varepsilon(\chi)\ E(\chi)\ V(\chi) d\chi, \tag{5}$$

where $\varepsilon(\chi)$, $E(\chi)$, and $V(\chi)$ are the lattice strain, the diffraction elastic constant, and the volume fraction for χ angle, respectively.

Figure 6a shows the $E^{hk.l}$ values as a function of χ angle. The whole χ angle region is, however, not possible to be covered by the *hk.l* peaks possibly analyzed with high accuracy. To estimate the $E(\chi)$ for the whole χ angle region, a cubic spline interpolation with 91 destinations is then employed. The $E(\chi)$ values are in a range of approximately 45 GPa, very close to the values reported by Gong et al. [19] and the values calculated according to the Kroner model [46] using the elastic constants for pure Mg reported by Slutsky and Garland (c_{11} = 59.40 GPa, c_{12} = 25.61 GPa, c_{13} = 21.44 GPa, c_{33} = 61.60 GPa, c_{44} = 16.40 GPa, and c/a = 1.623) [47]. The anisotropy in the diffraction elastic constant is not very large, i.e., the values are almost similar for different [*hk.l*] oriented grains. Figure 6b shows the $\varepsilon(\chi)$ values for some steps of applied true strains. To estimate the $\varepsilon(\chi)$ values for the whole χ angle region, the cubic spline interpolation with 91 destinations is also employed. Figure 6c shows the intensity of the basal pole of HCP, which is a plot of the $I(\chi)$ value as a function of χ angle. This plot is often called an angle distribution function (ADF). The ADF is a cut through a line from the center to the perimeter in the pole figure [19,48,49], which was extracted from the orientation distribution function (ODF) using the MTEX program [50]. The ODF was determined in the Rietveld-texture analysis [32,33] of each diffraction pattern, assuming a fiber texture using the twelfth-order spherical harmonic function implemented in the MAUD software, v. 2.78 [34]. The $V(\chi)$ was then determined from the ADF with the following equation [49], and the results are shown in Figure 6d.

$$V(\chi) = I(\chi)\ \sin\chi\ \Delta\chi. \tag{6}$$

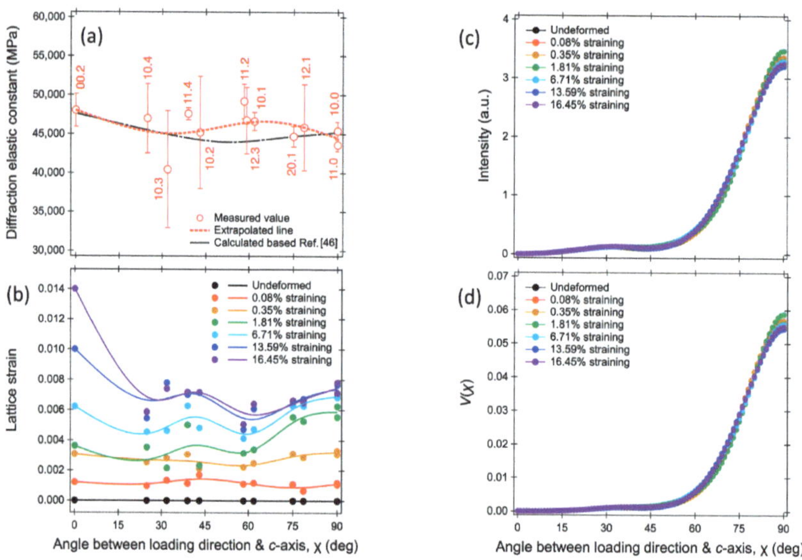

Figure 6. The parameters used in the newly proposed procedure for stress evaluation: (**a**) Diffraction elastic constants of the [*hk.l*] oriented grains versus χ (the angle between the *c*-axis and LD). (**b**) Lattice strains of the [*hk.l*] oriented grains versus χ for some applied true strain steps. (**c**) The axial distribution function (ADF) curves at some steps of applied true strains. (**d**) The volume fraction of [*hk.l*] grains versus χ for some applied true strain steps.

The phase stress values evaluated using our new proposed method are shown in Figure 7, which are in very good agreement with the applied true stress values not only for the elastic regime but for the whole deformation. These results validate the high accuracy of our newly proposed procedure to evaluate stress values using multiple *hk.l* peaks. This result indicates also that the lattice strains for the LD are enough and reliable to evaluate the macroscopic stress in the uniaxial loading experiments. In uniaxial loading experiments, there are two principal stress tensors parallel to the LD and TD. However, the [*hkl*]//TD grains do not have the same [*hkl*] orientations with grains having the scattering vector parallel to the LD; instead, they include various [*hkl*] orientations. Moreover, the orthogonal directions of [*hkl*]//TD grains are not always parallel to the LD. This issue has already been mentioned in several papers [51–53]. Stress conversions using the lattice strains for the TD without careful consideration of the development of texture in the LD and TD are, therefore, messy and require caution. The use of a simulation using the EPSC modelling [17,35,37,40] could be useful to understand the lattice strain behavior in the TD but is very difficult if the specimen is multiphase or undergoes phase transformation during deformation like in TRIP steels. Our new proposed method will be helpful to understand the individual average stresses of the constituent phases in multiphase materials with or without phase transformation.

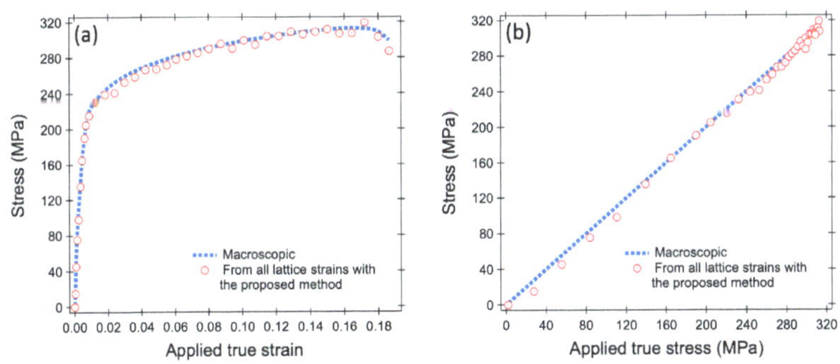

Figure 7. Average phase stress values evaluated using the newly proposed procedure versus (**a**) the applied true strains and (**b**) the applied true stresses.

On the other hand, the engineering parts, such as welded and plastically formed parts, most of which are machined under multi-axial loading, require several lattice strain tensors to determine the residual stresses. When measuring the residual stresses in the engineering components made of Mg alloys, the following procedure is considered applicable, considering the isotropic diffraction elastic constant. For simplicity, a triaxial principal stress condition is assumed.

$$\varepsilon_{ii}^{\text{ave III}} = \int_{\chi=0^\circ}^{\chi=90^\circ} \varepsilon_{ii}(\chi)\, V_{ii}(\chi) d\chi, \tag{7}$$

where *ii* is the principal stress direction, 11, 22, or 33. Equation (7) is the integral of the lattice strains weighted by the [*hk.l*] grain volume fractions and, henceforth, referred to as the average III phase strain ($\varepsilon^{\text{ave III}}$). The residual stresses can then be evaluated from the $\varepsilon_{ii}^{\text{ave III}}$ values according to the following equation:

$$\sigma_{ii} = \frac{E}{1+\nu}\left[\varepsilon_{ii}^{\text{ave III}} + \frac{\nu}{1-2\nu}\left(\varepsilon_{11}^{\text{ave III}} + \varepsilon_{22}^{\text{ave III}} + \varepsilon_{33}^{\text{ave III}}\right)\right]. \tag{8}$$

Here, *E* is Young's modulus, and ν is Poisson's ratio. The $\varepsilon^{\text{ave III}}$ values for the LD calculated with Equation (7) are plotted in Figure 4 and show a good linear relationship with the applied true stress.

3.4. The Lattice Strain of Particular Grain for Phase Stress

Figure 8a shows two $\varepsilon^{hk.l}$ values having linear responses to the applied true stress for the whole deformation: the [12.1] and [11.4] oriented grains in the LD. The relative integrated peak intensity results for 12.1 and 11.4 (shown in Figure 3a′,b′) are replotted in Figure 8b,c. The relative integrated peak intensity of 11.4, though the scatter is large due to the low intensity before deformation, decreases with the plastic deformation in the LD (Figure 8b), while it shows little change in the TD (Figure 8c). The decrease in the relative integrated peak intensity of 11.4 in the LD may be because that peak became very broad, and the profile fitting was difficult to perform properly. However, it is very difficult to determine that the integrated peak intensity of 11.4 for both the LD and TD has hardly changed. Moreover, the integrated peak intensity of 11.4 in the LD before deformation is very low due to the strong extrusion texture (Figure 2a). On the contrary, the relative integrated intensities of 12.1 for the LD and TD are consistently showing slight changes during deformation. It can also be observed in Figure 2a that the integrated peak intensities of 12.1 for the LD and TD before deformation are close to each other, even though the sample was extruded. The [12.1] orientation may be insensitive to the texture development.

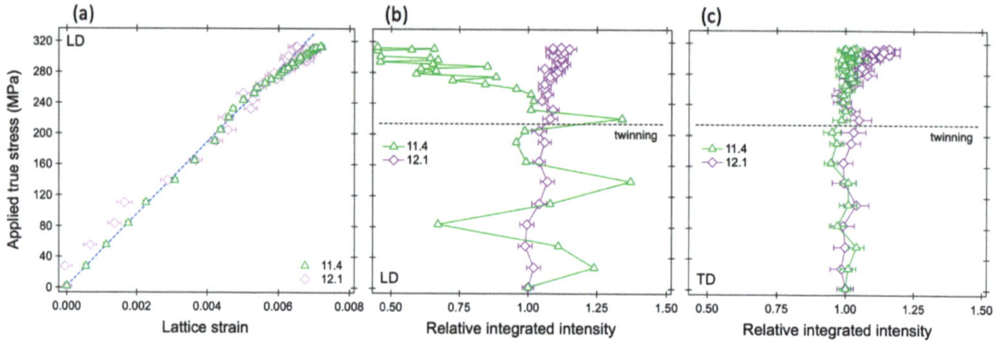

Figure 8. (**a**) Lattice strains of the [12.1] and [11.4] oriented grains in the loading direction. Relative integrated intensities of the 12.1 and 11.4 peaks in (**b**) the loading direction and (**c**) the transverse direction.

To understand the reason, the values of the Schmid factor for three deformation modes during tension, basal slip, extension twinning, and prismatic slip, were calculated. The calculations for basal slip or prismatic slip were conducted on three variants, and the maximum values were adopted. The calculations for extension twinning were conducted on six variants, and the maximum positive values were adopted. The variants used for the calculation refer to Ref. [19]. Figure 9 shows the distributions of the values of the Schmid factor as functions of χ angle. The value of the Schmid factor for basal slip is zero at the χ angle of zero, increases with increasing χ angle to reach a maximum value at approximately the χ angle of 45°, and then decreases toward zero at the χ angle of 90°. The value of the Schmid factor for extension twinning is maximum at the χ angle of zero and decreases toward zero at the χ angle of 90°, and it is vice versa for prismatic slip.

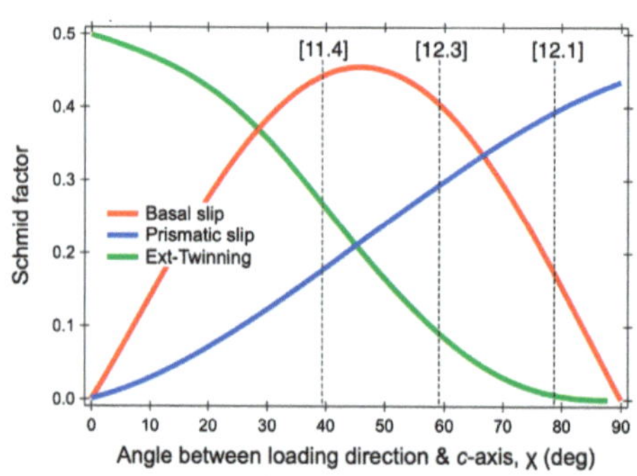

Figure 9. Distributions of values of the Schmid factor for basal slip, extension twinning, and prismatic slip as functions of χ angle.

The [12.3] oriented grains show a large value of the Schmid factor for basal slip, explaining that the earlier deviation from the applied stress shown in Figure 3c is due to the basal slip. Therefore, the accuracy of stress measurement using the 12.3 peak should receive attention for the evaluation of residual stresses in the HCP-structured engineering parts after plastic deformation. The [11.4] oriented grains also have a very large value of the Schmid factor for basal slip, being larger than the values of the Schmid factor for extension twinning and prismatic slip. The linear relationship between the $\varepsilon^{hk.l}$ value for [11.4] grains and the applied true stress for the whole deformation shown in Figure 8 may, therefore, have been a contingent one. Meanwhile, the Schmid factor of [12.1] oriented grains shows a large value for prismatic slip, being larger than that for basal slip, and that for extension twinning is almost zero. The critical resolving shear stress (CRSS) value is also an important parameter for determining the activation of the deformation mode. The CRSS value is smallest for basal slip, increasing in order of extension twinning and then prismatic slip. The CRSS for prismatic slip is generally three times larger than that for basal slip [54]. Muransky et al. [17] performed a simulation using the EPSC modelling to describe the responses of $\varepsilon^{hk.l}$ values to the applied stress measured with neutron diffraction for an AZ31 alloy and found also that the CRSS value for prismatic slip (~90 MPa) is almost three times larger than that for basal slip (~30 MPa). Thus, we can expect that, in the [12.1] oriented grains, prismatic slip occurs at the later stage of deformation. Similar EPSC modelling could be performed to understand the detailed deformation modes, but since there have already been many reports on this topic [17,35,37,40], we omit it. However, based on these discussions, we newly recommend that the [12.1] orientation is suitable for stress evaluation for HCP-structured Mg alloys not only in the neutron diffraction stress measurement but also in the X-ray diffraction one when multiple [*hk.l*] orientations cannot be obtained simultaneously. We believe that the [12.1] orientation is also valid for other HCP-structured materials, but clarification is still needed to be performed in the future.

4. Conclusions

In this study, tensile deformation in situ neutron diffraction of an extruded AZ31 alloy was performed to monitor the crystallographic evolutions as deformation progresses. The conventional procedures to evaluate stress values from the lattice strain values for HCP-structured materials were reviewed and compared to the stress values obtained mechanically. We also propose a new procedure to evaluate the stress values that represent macroscopically or as the phase average value. The results are summarized as follows.

1. The increases in the lattice strains with respect to the applied true stress after yielding largely vary among [*hk.l*] grains. Some [*hk.l*] grains have little or no increase in lattice strain, making it difficult to use the conventional procedures for determining the average phase strain by using lattice constants or by averaging several lattice strains.
2. Our newly proposed procedure of stress evaluation from the lattice strains shows very high accuracy and reliability by weighting the volume fraction of [*hk.l*] grains and evaluating them in many [*hk.l*] orientations in addition to multiplication by the diffraction elastic constant.
3. The conventionally recommended 12.3 peak for stress measurement of HCP-structured materials should be used with caution for the evaluation of stresses of HCP-structured engineering parts due to the huge stress relaxation in 12.3 oriented grains by basal slip.
4. We recommend to use the 12.1 peak for stress evaluation when multiple *hk.l* peaks cannot be obtained simultaneously. The lattice strain value evaluated from the 12.1 peak shows a good linear relationship with the applied true stress for the whole deformation region.

Author Contributions: Conceptualization, S.H., W.G. and T.K.; formal analysis, S.H.; data curation, S.H. and W.G.; writing—original draft preparation, S.H.; writing—review and editing, S.H., W.G. and T.K. All authors have read and agreed to the published version of the manuscript.

Funding: This study was financially supported by the KAKENHI (JP18H05479 and JP18H05476), JST-CREST (JPMJCR2094), and MEXT Program: Data Creation and Utilization Type Material Research and Development (JPMXP1122684766), all through the Ministry of Education, Culture, Sports, Science and Technology (MEXT), Japan.

Data Availability Statement: The data presented in this study are available on request from the corresponding author.

Acknowledgments: The authors acknowledge Michiaki Yamasaki for providing the sample used in this study. The neutron diffraction experiments were performed at TAKUMI in MLF of J-PARC with the proposals of 2021P0400 and 2021I0019.

Conflicts of Interest: The authors declare no conflict of interest.

References

1. Joost, W.J.; Krajewski, P.E. Towards magnesium alloys for high-volume automotive applications. *Scr. Mater.* **2017**, *128*, 107–112. [CrossRef]
2. Agnew, S.R. Wrought magnesium: A 21st century outlook. *JOM* **2004**, *56*, 20–21. [CrossRef]
3. Kawamura, Y.; Hayashi, K.; Inoue, A.; Masumoto, T. Rapidly Solidified Powder Metallurgy $Mg_{97}Zn_1Y_2$Alloys with Excellent Tensile Yield Strength above 600 MPa. *Mater. Trans.* **2001**, *42*, 1172–1176. [CrossRef]
4. Blum, W.; Zhang, P.; Watzinger, B.; Grossmann, B.V.; Haldenwanger, H.G. Comparative study of creep of the die-cast Mg-alloys AZ91, AS21, AS41, AM60, and AE42. *Mater. Sci. Eng. A* **2001**, *319–321*, 735–740. [CrossRef]
5. Itoi, T.; Inazawa, T.; Yamasaki, M.; Kawamura, Y.; Hirohashi, M. Microstructure and mechanical properties of Mg-Zn-Y alloy sheet prepared by hot-rolling. *Mater. Sci. Eng. A* **2013**, *560*, 216–223. [CrossRef]
6. Noyan, I.C. Equilibrium Conditions for the Average Stresses Measured By X-rays. *Metall. Trans. A Phys. Metall. Mater. Sci.* **1983**, *14*, 1907–1914. [CrossRef]
7. Harjo, S.; Tomota, Y.; Ono, M. Measurements of thermal residual elastic strains in ferrite–austenite Fe–Cr–Ni alloys by neutron and X-ray diffractions. *Acta Mater.* **1998**, *47*, 353–362. [CrossRef]
8. Lin, L.; Liu, Z.; Zhuang, W.; Peng, H. Effects of pre-strain on the surface residual stress and corrosion behavior of an Al-Zn-Mg-Cu alloy plate. *Mater. Charact.* **2020**, *160*, 110129. [CrossRef]
9. Hutchings, M.T. Neutron diffraction measurement of residual stress fields-the engineer's dream come true? *Neutron News* **1992**, *3*, 14–19. [CrossRef]
10. Nishida, S.; Nishino, S.; Sekine, M.; Oka, Y.; Harjo, S.; Kawasaki, T.; Suzuki, H.; Morii, Y.; Ishii, Y. Analysis of Residual Stress in Steel Bar Processed by Cold Drawing and Straightening. *Mater. Trans.* **2021**, *62*, 667–674. [CrossRef]
11. Harjo, S.; Tsuchida, N.; Abe, J.; Gong, W. Martensite phase stress and the strengthening mechanism in TRIP steel by neutron diffraction. *Sci. Rep.* **2017**, *7*, 15149. [CrossRef]
12. Harjo, S.; Gong, W.; Aizawa, K.; Kawasaki, T.; Yamasaki, M. Strengthening of αMg and long-period stacking ordered phases in a Mg-Zn-Y alloy by hot-extrusion with low extrusion ratio. *Acta Mater.* **2023**, *255*, 119029. [CrossRef]
13. Withers, P.J.; Bhadeshia, H.K.D.H. Residual stress. Part 1—Measurement techniques. *Mater. Sci. Technol.* **2001**, *17*, 355–365. [CrossRef]
14. Koyanagi, Y.; Ueta, S.; Kawasaki, T.; Harjo, S.; Cho, K.; Yasuda, H.Y.H.Y. Investigation of strengthening mechanism in Ni–38Cr-3.8Al alloy with fine lamellar structure by in situ neutron diffraction analysis. *Mater. Sci. Eng. A* **2020**, *773*, 138822. [CrossRef]
15. Daymond, M.R.; Bourke, M.A.M.; Von Dreele, R.B.; Clausen, B.; Lorentzen, T. Use of Rietveld refinement for elastic macrostrain determination and for evaluation of plastic strain history from diffraction spectra. *J. Appl. Phys.* **1997**, *82*, 1554–1562. [CrossRef]
16. Harjo, S.; Kawasaki, T.; Grazzi, F.; Shinohara, T.; Tanaka, M. Neutron diffraction study on full-shape Japanese sword. *Materialia* **2019**, *7*, 100377. [CrossRef]
17. Muránsky, O.; Carr, D.G.; Barnett, M.R.; Oliver, E.C.; Šittner, P. Investigation of deformation mechanisms involved in the plasticity of AZ31 Mg alloy: In situ neutron diffraction and EPSC modelling. *Mater. Sci. Eng. A* **2008**, *496*, 14–24. [CrossRef]
18. Muránsky, O.; Barnett, M.R.; Carr, D.G.; Vogel, S.C.; Oliver, E.C. Investigation of deformation twinning in a fine-grained and coarse-grained ZM20 Mg alloy: Combined in situ neutron diffraction and acoustic emission. *Acta Mater.* **2010**, *58*, 1503–1517. [CrossRef]
19. Gong, W.; Aizawa, K.; Harjo, S.; Zheng, R.; Kawasaki, T.; Abe, J.; Kamiyama, T.; Tsuji, N. Deformation behavior of as-cast and as-extruded Mg97Zn1Y2 alloys during compression, as tracked by in situ neutron diffraction. *Int. J. Plast.* **2018**, *111*, 288–306. [CrossRef]
20. Woo, W.; Choo, H.; Prime, M.B.; Feng, Z.; Clausen, B. Microstructure, texture and residual stress in a friction-stir-processed AZ31B magnesium alloy. *Acta Mater.* **2008**, *56*, 1701–1711. [CrossRef]
21. Turski, M.; Paradowska, A.; Zhang, S.Y.; Mortensen, D.; Fjaer, H.; Grandfield, J.; Davis, B.; Delorme, R. Validation of predicted residual stresses within direct chill cast magnesium alloy slab. *Metall. Mater. Trans. A Phys. Metall. Mater. Sci.* **2012**, *43*, 1547–1557. [CrossRef]
22. Brown, D.W.; Agnew, S.R.; Bourke, M.A.M.; Holden, T.M.; Vogel, S.C.; Tomé, C.N. Internal strain and texture evolution during deformation twinning in magnesium. *Mater. Sci. Eng. A* **2005**, *399*, 1–12. [CrossRef]

23. Prevey, P.S. X-Ray Diffraction Residual Stress Techniques. In *Materials Characterization*; ASM International: Almere, The Netherlands, 1986; pp. 380–392. [CrossRef]
24. Harjo, S.; Ito, T.; Aizawa, K.; Arima, H.; Abe, J.; Moriai, A.; Iwahashi, T.; Kamiyama, T. Current Status of Engineering Materials Diffractometer at J-PARC. *Mater. Sci. Forum.* **2011**, *681*, 443–448. [CrossRef]
25. Nakajima, K.; Kawakita, Y.; Itoh, S.; Abe, J.; Aizawa, K.; Aoki, H.; Endo, H.; Fujita, M.; Funakoshi, K.; Gong, W.; et al. Materials and Life Science Experimental Facility (MLF) at the Japan Proton Accelerator Research Complex II: Neutron Scattering Instruments. *Quantum Beam Sci.* **2017**, *1*, 9. [CrossRef]
26. Máthis, K.; Drozdenko, D.; Németh, G.; Harjo, S.; Gong, W.; Aizawa, K.; Yamasaki, M.; Kawamura, Y. In-situ Investigation of the Microstructure Evolution in Long-Period-Stacking-Ordered (LPSO) Magnesium Alloys as a Function of the Temperature. *Front. Mater.* **2019**, *6*, 270. [CrossRef]
27. Harjo, S.; Aizawa, K.; Gong, W.; Kawasaki, T. Neutron diffraction monitoring of as-cast Mg97Zn1Y2 during compression and tension. *Mater. Trans.* **2020**, *61*, 828–832. [CrossRef]
28. Ito, T.; Nakatani, T.; Harjo, S.; Arima, H.; Abe, J.; Aizawa, K.; Moriai, A. Application Software Development for the Engineering Materials Diffractometer, TAKUMI. *Mater. Sci. Forum.* **2010**, *652*, 238–242. [CrossRef]
29. Kawasaki, T.; Nakamura, T.; Gong, W.; Oikawa, K. Testing of time-of-flight neutron diffraction imaging using a high-resolution scintillator detector with wavelength-shifting fibers. *Phys. B Condens. Matter* **2018**, *551*, 460–463. [CrossRef]
30. Mao, W.; Gong, W.; Kawasaki, T.; Harjo, S. Effect of deformation-induced martensitic transformation on nonuniform deformation of metastable austenitic steel. *Mater. Sci. Eng. A* **2022**, *837*, 142758. [CrossRef]
31. Oishi-Tomiyasu, R.; Yonemura, M.; Morishima, T.; Hoshikawa, A.; Torii, S.; Ishigaki, T.; Kamiyama, T. Application of matrix decomposition algorithms for singular matrices to the Pawley method in Z-Rietveld. *J. Appl. Crystallogr.* **2012**, *45*, 299–308. [CrossRef]
32. Xu, P.; Harjo, S.; Ojima, M.; Suzuki, H.; Ito, T.; Gong, W.W.; Vogel, S.C.; Inoue, J.; Tomota, Y.Y.; Aizawa, K.; et al. High stereographic resolution texture and residual stress evaluation using time-of-flight neutron diffraction. *J. Appl. Crystallogr.* **2018**, *51*, 746–760. [CrossRef]
33. Wenk, H.-R.; Lutterotti, L.; Vogel, S.C. Rietveld texture analysis from TOF neutron diffraction data. *Powder Diffr.* **2010**, *25*, 283–296. [CrossRef]
34. Lutterotti, L.; Matthies, S.; Wenk, H.-R.; Schultz, A.S.; Richardson, J.W. Combined texture and structure analysis of deformed limestone from time-of-flight neutron diffraction spectra. *J. Appl. Phys.* **1997**, *81*, 594–600. [CrossRef]
35. Gong, W.; Zheng, R.; Harjo, S.; Kawasaki, T.; Aizawa, K.; Tsuji, N. In-situ observation of twinning and detwinning in AZ31 alloy. *J. Magnes. Alloy.* **2022**, *10*, 3418–3432. [CrossRef]
36. Harjo, S.; Gong, W.; Aizawa, K.; Kawasaki, T.; Yamasaki, M.; Mayama, T.; Kawamura, Y. Effect of Extrusion Ratio in Hot-Extrusion on Kink Deformation during Compressive Deformation in an αMg/LPSO Dual-Phase Magnesium Alloy Monitored by In Situ Neutron Diffraction. *Mater. Trans.* **2023**, *64*, 766–773. [CrossRef]
37. Čapek, J.; Máthis, K.; Clausen, B.; Stráská, J.; Beran, P.; Lukáš, P. Study of the loading mode dependence of the twinning in random textured cast magnesium by acoustic emission and neutron diffraction methods. *Mater. Sci. Eng. A* **2014**, *602*, 25–32. [CrossRef]
38. Agnew, S.R. *Deformation Mechanisms of Magnesium Alloys*; Woodhead Publishing Limited: Thorston, UK, 2012. [CrossRef]
39. Liu, M.; Diercks, P.; Manzoni, A.; Čížek, J.; Ramamurty, U.; Banhart, J. Positron annihilation investigation of thermal cycling induced martensitic transformation in NiTi shape memory alloy. *Acta Mater.* **2021**, *220*, 117298. [CrossRef]
40. Barnett, M.R. Twinning and the ductility of magnesium alloys. Part I: "Tension" twins. *Mater. Sci. Eng. A* **2007**, *464*, 1–7. [CrossRef]
41. Pawley, G.S. Unit-cell refinement from powder diffraction scans. *J. Appl. Crystallogr.* **1981**, *14*, 357–361. [CrossRef]
42. Rietveld, H.M. Line profiles of neutron powder-diffraction peaks for structure refinement. *Acta Crystallogr.* **1967**, *22*, 151–152. [CrossRef]
43. Asoo, K.; Tomota, Y.Y.; Harjo, S.; Okitsu, Y. Tensile behavior of a trip-aided ultra-fine grained steel studied by neutron diffraction. *ISIJ Int.* **2011**, *51*, 145–150. [CrossRef]
44. Fukuda, K.; Tomota, Y.; Harjo, S.; Gong, W.; Woo, W.; Seong, B.S.S.; Kuwahara, Y.; Ikuta, F. Residual stresses in steel rod with collar formed by partial diameter-enlarging technique. *Mater. Sci. Technol.* **2017**, *33*, 172–180. [CrossRef]
45. Paddea, S.; Francis, J.A.; Paradowska, A.M.; Bouchard, P.J.; Shibli, I.A. Residual stress distributions in a P91 steel-pipe girth weld before and after post weld heat treatment. *Mater. Sci. Eng. A* **2012**, *534*, 663–672. [CrossRef]
46. Kröner, E. Berechnung der elastischen Konstanten des Vielkristalls aus den Konstanten des Einkristalls. *Z. Phys.* **1958**, *151*, 504–518. [CrossRef]
47. Slutsky, L.J.; Garland, C.W. Elastic constants of magnesium from 4.2°K to 300°K. *Phys. Rev.* **1957**, *107*, 972–976. [CrossRef]
48. Clausen, B.; Tomé, C.N.; Brown, D.W.; Agnew, S.R. Reorientation and stress relaxation due to twinning: Modeling and experimental characterization for Mg. *Acta Mater.* **2008**, *56*, 2456–2468. [CrossRef]
49. Agnew, S.R.; Mulay, R.P.; Polesak, F.J.; Calhoun, C.A.; Bhattacharyya, J.J.; Clausen, B. In situ neutron diffraction and polycrystal plasticity modeling of a Mg-Y-Nd-Zr alloy: Effects of precipitation on individual deformation mechanisms. *Acta Mater.* **2013**, *61*, 3769–3780. [CrossRef]

50. Hielscher, R.; Schaeben, H. A novel pole figure inversion method: Specification of the MTEX algorithm. *J. Appl. Crystallogr.* **2008**, *41*, 1024–1037. [CrossRef]
51. Ungár, T.; Stoica, A.D.; Tichy, G.; Wang, X.L. Orientation-dependent evolution of the dislocation density in grain populations with different crystallographic orientations relative to the tensile axis in a polycrystalline aggregate of stainless steel. *Acta Mater.* **2014**, *66*, 251–261. [CrossRef]
52. Xie, Q.; Yan, Z.; Yu, D.; An, K.; Yan, X.; Yin, S.; Gillham, B.; Wu, X.; Yang, P.; Zhao, Z.; et al. Crystallographic orientation and spatially resolved damage for polycrystalline deformation of a high manganese steel. *Acta Mater.* **2022**, *226*, 117628. [CrossRef]
53. Tomota, Y.; Ojima, M.; Harjo, S.; Gong, W.; Sato, S.; Ungár, T. Dislocation densities and intergranular stresses of plastically deformed austenitic steels. *Mater. Sci. Eng. A* **2019**, *743*, 32–39. [CrossRef]
54. Yu, M.; Cui, Y.; Wang, J.; Chen, Y.; Ding, Z.; Ying, T.; Llorca, J.; Zeng, X. Critical resolved shear stresses for slip and twinning in Mg-Y-Ca alloys and their effect on the ductility. *Int. J. Plast.* **2023**, *162*, 103525. [CrossRef]

quantum beam science

MDPI

Article

Principal Preferred Orientation Evaluation of Steel Materials Using Time-of-Flight Neutron Diffraction

Pingguang Xu [1,*], Shuyan Zhang [2,†], Stefanus Harjo [3], Sven C. Vogel [4] and Yo Tomota [1,5]

1 Materials Sciences Research Center, Japan Atomic Energy Agency, Tokai 319-1195, Ibaraki, Japan
2 ISIS Facility, Rutherford Appleton Laboratory, Harwell Science and Innovation Campus, Didcot OX11 0QX, UK
3 J-PARC Center, Japan Atomic Energy Agency, Tokai 319-1195, Ibaraki, Japan
4 Los Alamos Neutron Science Center (LANSCE), Los Alamos National Laboratory, Los Alamos, NM 87545, USA
5 Graduate School of Science and Engineering, Ibaraki University, Hitachi 316-8511, Ibaraki, Japan
* Correspondence: xu.pingguang@jaea.go.jp; Tel.: +81-29-282-6885
† Current address: Centre of Excellence for Advanced Materials, Dongguan 523808, Guangdong, China.

Abstract: Comprehensive information on in situ microstructural and crystallographic changes during the preparation/manufacturing processes of various materials is highly necessary to precisely control the microstructural morphology and the preferred orientation (or texture) characteristics for achieving an excellent strength–ductility–toughness balance in advanced engineering materials. In this study, in situ isothermal annealing experiments with cold-rolled 17Ni-0.2C (mass%) martensitic steel sheets were carried out by using the TAKUMI and ENGIN-X time-of-flight neutron diffractometers. The inverse pole figures based on full-profile refinement were extracted to roughly evaluate the preferred orientation features along three principal sample directions of the investigated steel sheets, using the General Structure Analysis System (GSAS) software with built-in generalized spherical harmonic functions. The consistent rolling direction (RD) inverse pole figures from TAKUMI and ENGIN-X confirmed that the time-of-flight neutron diffraction has high repeatability and statistical reliability, revealing that the principal preferred orientation evaluation of steel materials can be realized through 90° TD → ND (transverse direction → normal direction) rotation of the investigated specimen on the sample stage during two neutron diffraction experiments. Moreover, these RD, TD, and ND inverse pole figures before and after the in situ experiments were compared with the corresponding inverse pole figures recalculated from the MUSASI-L complete pole figure measurement and the HIPPO in situ microstructure evaluation, respectively. The similar orientation distribution characteristics suggested that the principal preferred orientation evaluation method can be applied to the in situ microstructural evolution of bulk orthorhombic materials and spatially resolved principal preferred orientation mappings of large engineering structure parts.

Keywords: neutron diffraction; preferred orientation; inverse pole figure; full-profile refinement; bulk texture; multiphase materials

1. Introduction

It is well known that the material properties of polycrystalline solids depend on their single-crystal anisotropic properties and the orientation distribution of the single crystals in the polycrystalline aggregate. Accurate texture measurements can help us to derive and optimize the process history of various thermomechanical treatments [1,2]. Considering that engineering materials usually involve inhomogeneous microstructural characteristics [1,2] due to the heterogeneous deformation and/or the steep temperature gradients and/or composition gradients, bulk texture measurement [3] and in situ preferred orientation evaluation [4–7] are valuable for elucidating the microstructure/texture evolution process, as well as for achieving the desired strength–ductility–toughness balance.

There were two angle-dispersive neutron diffractometers RESA-2 and MUSASI-L at Japan Research Reactor No.3 (JRR-3)'s guide hall for bulk texture evaluations of metallic materials. However, unlike the recently upgraded WOMBAT neutron diffractometer, with a wide-angle ($\Delta 2\theta = 120°$), large-area, curved, position-sensitive detector located at the Open-Pool Australian Lightwater Reactor (OPAL) guide hall [8], the pole figure measurement speeds of RESA-2 (with a narrow one-dimensional position-sensitive detector) and MUSASI-L (with a single-tube detector) are very slow and not suitable for in situ time-sliced texture measurements. In view of the new era of stress and texture evaluation techniques involved in various neutron diffraction instruments from the steady-state reactor neutron source at JRR-3, the large spallation neutron source at J-PARC, and the compact neutron source at RANS, these RESA-2 and MUSASI-L instruments have been replaced by the residual stress analyzer RESA at the T2-1 neutron beamline port of JRR-3 in order to improve the cost performance.

To date, the time-of-flight neutron diffraction technique has been widely employed to investigate the hot/warm/room-temperature/cryogenic elastoplastic deformation, transformation, and recrystallization behaviors of metallic materials [9–11], and several new neutron diffractometers have been established to meet such rapidly increasing needs [12–14]. As a new-generation engineering materials neutron diffractometer, TAKUMI adopts the event-type neutron data recording technique, enabling good neutron diffractogram conversion treatment of the experimental raw data using optimal time-slicing parameters [15] in real time, bridging the bulk microstructure changes and the macroscopic mechanical and physical response evolutions of advanced materials under various extreme conditions, and even identifying the deep relationships among the microstructures, preferred orientations, and mechanical anisotropic properties of complex engineering components through the proper combination of radial collimators and adjustable beam slits. Moreover, the orientation investigation is essential for clarifying the complex microstructural evolution processes involved in several competitive behaviors (e.g., recrystallization and precipitation during the formation process of ultrafine-grained multiphase microstructures).

Moreover, the Z-Rietveld software [16] with built-in March–Dollase functions [17] has been developed to analyze the powder diffraction data obtained from various diffractometers, applicable for both needle- and plate-shaped crystals, which is explicitly correct when the sample has cylindrical symmetry along the diffraction vector and a reasonable approximation when the cylinder's axis is perpendicular to the diffraction plane [18]. However, the generalized harmonic spherical function [19] has not been incorporated in this software to deal with the more complex preferred orientations during the full neutron diffractogram refinement. Considering that the General Structure Analysis System (GSAS) software [20] with the above two built-in functions is widely used, its application reliability for principal preferred orientation evaluation is worth examining for wider application to various in situ neutron diffraction experiments and principal preferred orientation mappings of large engineering structure parts.

In this study, the in situ neutron diffraction event-type and histogram-type diffractograms of a cold-rolled 17Ni-0.2C (mass%) martensitic steel, as a typical textured material, were acquired from the TAKUMI and ENGIN-X neutron diffractometers, respectively, and the converted histogram-type neutron diffractograms were refined using the GSAS software. The inverse pole figures were extracted to roughly evaluate the preferred orientation features along the triaxial directions of the investigated steel sheets. The complete pole figures before and after the isothermal annealing were measured using the MUSASI-L angle-dispersive neutron diffractometer, and the microstructural changes and texture evolution during the isothermal annealing were measured in situ using the HIPPO time-of-flight neutron diffractometer to further confirm the above technique's reliability. However, the in situ microstructural evolution will be discussed in detail in another paper on ultrafine-grained multiphase microstructures to clarify the competitive behavior between the static recrystallization of cold-rolled martensite and the static precipitation of austenite during the isothermal annealing.

2. Neutron Instruments and Analysis Method

2.1. Neutron Instruments

The TAKUMI engineering materials neutron diffractometer at J-PARC is equipped with 3600 channels of scintillator detectors with spatial resolutions of 3.0 mm horizontally and 200 mm vertically [21], located at L_2 = 2.00 m from the sample center (the secondary flightpath). The north bank ($2\theta = -90°$ scattering bank) and the south bank ($2\theta = +90°$ scattering bank) cover a 2θ range of $-105°{\sim}-75°$ and 75°~105°, respectively, at a vertical angle η range of $-16°{\sim}16°$. The distance from the neutron source target to the sample stage center (the primary flightpath, L_1) is 40.00 m, and the pulsed neutron beam is supplied at 25 Hz for a high-intensity neutron beam mode (or 12.5 Hz for a high-resolution neutron beam mode), which enabled us to obtain the neutron diffractogram corresponding to a lattice spacing distance d range of 0.5~2.5 Å (or 0.5~5.0 Å). Figure 1 shows an example sample environment setup of the TAKUMI instrument for in situ isothermal annealing experiments using an infrared furnace.

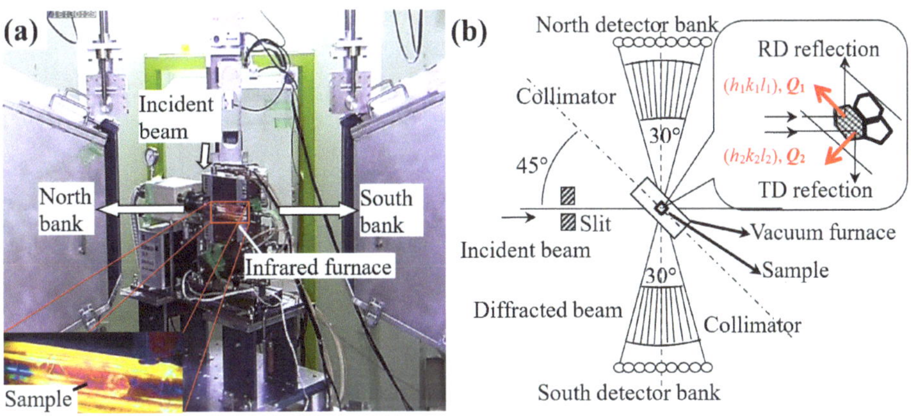

Figure 1. (**a**) General view and (**b**) instrumental distribution of the TAKUMI time-of-flight neutron diffractometer during in situ neutron diffraction measurements.

ENGIN-X is a well-known time-of-flight neutron diffractometer located at the ISIS facility, with the flight paths L_1 = 50.0 m and L_2 = 1.50 m [22]. Its north and south banks are built with ZnS scintillators in a 3 mm horizontal resolution, covering a 2θ range of 76°~104° and a vertical angle range of $-21°{\sim}21°$. It was built for non-destructive stress evaluation in the research fields of materials science and engineering, including the new welding technologies for airframe manufacturing, the fatigue crack initiation and propagation in composite materials, the thermal cycling of materials used in the power generation industry, the development of strain measurement standards, and so on. It has been also employed to study the microstructural evolution during various thermomechanical controlled processes by using the GSAS full-diffractogram refinement technique.

Unlike the above two engineering materials neutron diffractometers, which have various radial collimators for reliably collecting the local neutron diffraction information of large semi-finished products, the HIPPO neutron diffractometer [23] is a well-known high-intensity powder diffractometer equipped with neutron diffraction detector panels for 2θ = 145, 90, 40° three-dimensional diffraction rings (about 1360 ^{3}He tubes) (note: in 2012, neutron detector panels for 2θ = 120 and 60° diffraction rings were added to the upgraded HIPPO to achieve a better stereographic angle coverage with reliable instrumental resolution $\Delta d/d$), mostly applying to in situ time-sliced high-pressure and/or crystallographic texture evolutions of various advanced materials.

2.2. Analysis Method

In general, for an arbitrary crystalline grain *i* in the gauge volume (Figure 1b), there were two lattice planes $(h_n k_n l_n)_i$ and $(h_s k_s l_s)_i$, perpendicular to one another, to diffract the incident neutrons forward to the north and south banks, respectively, because their diffraction vectors Q_1 and Q_2 were about 45° from the incident beam direction in the horizontal plane. For all of the crystalline grains (i = 1, 2, 3......) in the gauge volume, these lattice planes $(h_n k_n l_n)_i$ and $(h_s k_s l_s)_i$ changed from the low (*hkl*)-index planes to the high (*hkl*)-index planes, with a specific diffraction intensity ratio related to the preferred orientation characteristics of the investigated specimen, as well as to the multiplicity of (*hkl*)-index reflection, the structure factor, and the Debye–Waller temperature factor [18–20]. All of the diffracted neutrons acquired by the two neutron detector banks finally formed the north-bank and south-bank neutron diffractograms, respectively. It should be mentioned that some diffracted neutrons from the high (*hkl*)-index lattice planes with a lattice plane spacing smaller than d_{c1} were not easy to distinguish because of the low diffraction intensities and the highly overlapping peaks. In addition, the frequency of neutron pulses also resulted in a limitation on the longest neutron flight time t_{c2}, i.e., there was a specific lattice plane $d_{c2} = h{\cdot}t_{c2}/(m{\cdot}(L_1 + L_2))$, where h was the Planck parameter and m was the neutron mass. Here, the crystallographic neutron diffraction characteristics of textured steel materials with a lattice plane spacing between d_{c1} = 0.5 and d_{c2} = 2.25 Å were usually taken into consideration.

During the neutron diffraction measurement, unlike the conventional histogram-type data acquisition system, the event-type data acquisition system records the relative time-of-flight neutron information for each detected neutron, including the spallation pulse information, the detector code number, and the position angle. On the base of the event-type neutron diffraction experimental data, the histogram-type neutron diffractograms are freely constructed after measurement [15]. If the obtained neutron diffractograms are not satisfactory, one may construct the histogram-type data again by simply changing the related parameters. The GSAS software [20] and the Materials Analysis Using Diffraction (MAUD) software [24] may be employed to analyze these time-of-flight neutron diffractograms through reading the TAKUMI instrumental parameter file and obtaining the analysis template file for further batch analysis, respectively.

Using the GSAS software together with a built-in generalized spherical harmonic function, the Rietveld method uses the least-squares approach to refine a theoretical line profile until it matches the measured profile [20]. The residual error $y_{io} - y_{ic}$ from the neutron intensities of the experimental observation (y_{io}) and the refinement calculation (y_{ic}) were comprehensively considered through the residual of least-squares refinement R_p and the weighted residual of least-squares refinement R_{wp}, calculated as follows [20]:

$$R_p = \frac{\sum|y_{io} - y_{ic}|}{\sum y_{io}} \tag{1}$$

$$R_{wp} = \left[\frac{\sum w_i{\cdot}(y_{io} - y_{ic})^2}{\sum w_i{\cdot}y_{io}^2}\right]^{\frac{1}{2}} \tag{2}$$

where w_i is the weighted coefficient dependent on the wavelength distribution of the incident neutron beam. The general criterion for an acceptable crystal structure diffraction analysis is $R_p \leq 0.10$ or $R_{wp} \leq 0.10$, where R_{wp} is only favorable when the higher-intensity data points are presumed to be more important than the lower-intensity data points.

3. Experimental Procedures

3.1. Sample Preparation

The chemical composition of the investigated steel was 17.2Ni–0.22C in mass%, which was employed in comparison with 18Ni steel for a fundamental investigation of the influence of the addition of carbon on the formation of ultrafine-grained multiphase microstructures toward a higher strength–plasticity–toughness balance. The steel was induction-melted, forged at 1100~900 °C, and groove-rolled into a 38 mm × 38 mm steel bar. A 65 mm × 22 mm × 3.6 mm steel block spark-cut from the volume center of the steel bar was solution-treated again at 1100 °C for 30 min, followed by water quenching and multi-pass cold rolling to achieve a 70% reduction in total thickness. The obtained 1.1 mm thick martensitic steel sheet was spark-cut into small specimens of 10 mm (rolling direction, RD) × 10 mm (transverse direction, TD) and stacked into cubes.

3.2. Experimental Procedures

During the TAKUMI neutron diffraction isothermal annealing experiment at 823 K, these cubic specimens were set up to enable their RDs along the nominal diffraction vector for the north-bank neutron detectors (Q_1) and their TDs along the nominal diffraction vector for the south-bank neutron detectors (Q_2). The pulsed neutron beam power was about 123 kW (note: the neutron beam power at J-PARC in 2023 is more than 800 kW) in the high-intensity neutron beam mode, covering a lattice spacing range of 0.5~2.5 Å; the incident beam size was 5 mmW × 8 mmH. Considering that the steel specimen was somewhat small, the average crystallographic information from a large gauge volume was necessary to improve the reliability of the orientation information in a short time, so the radial collimators at TAKUMI were not installed in these neutron diffraction experiments. The obtained preferred orientation parameters during the TAKUMI full-diffraction-profile refinement were extracted to draw the corresponding inverse pole figures.

During the similar ENGIN-X in situ neutron diffraction experiment for reference, the stacked specimens were set up through a 90° cubic sample rotation of TD → ND (transverse direction → normal direction) to enable their RDs along the nominal diffraction vector for the north-bank neutron detectors (Q_1) and their NDs (normal directions) along the nominal diffraction vector for the south-bank neutron detectors (Q_2). The pulsed neutron beam power was about 158 kW and the incident beam size was 5 mmW × 8 mmH during the in situ isothermal annealing experiment at 823 K. During the GSAS Rietveld full-diffractogram refinement [20], the rolling sample symmetry was employed, and the maximum expansion series of the spherical harmonic function was $L_{max} = 8$.

Before and after the TAKUMI and ENGIN-X in situ neutron diffractions, the ferrite (200) and (211) and the austenite (200) and (220) complete pole figures of these steel specimens were measured by using the MUSASI-L versatile angle-dispersive neutron diffractometer quipped with a pyrolytic graphite (PG) monochromator at the JRR-3 guide hall. According to the classic texture analysis theory [19], two complete pole figures from individual (non overlapping) diffraction peaks may provide enough crystallographic orientation information to calculate the orientation distribution function of engineering materials with a cubic crystal symmetry; the complete pole figure measurements of ferrite (110) and austenite (111) with overlapping peaks were avoided here. The 20 mmW × 20 mmH incident and diffraction neutron beams at a wavelength of $\lambda = 1.23$ Å were employed to ensure that the sample was completely bathed in them at each stereographic angle position. The distance between the φ50 mm (40 mm in the effective detect width) single-tube detector and the sample center was about 600 mm. The incident and diffracted collimators with a selective collimation angle $\alpha_3 = 0.33°$ and 0.67°, respectively, were employed. The LaboTex 3.0 texture software with a direct discrete ADC (arbitrary defined cells) method [25] was employed to calculate the inverse pole figures.

Furthermore, an in situ neutron diffraction experiment using the same 17Ni-0.2C (mass%) martensitic steel was carried out on HIPPO with step-by-step sample rotations of $\omega = 0$, 45, 67.5, and 90° to further confirm the principal preferred orientation evaluation through avoiding the possible sample misalignment. The Rietveld texture analysis was carried out using the MAUD software [24], using the generalized spherical harmonic function with the orthorhombic sample symmetry, and the same expansion series of spherical harmonic function $L_{max} = 8$ was employed here for reference.

4. Results and Discussion

Figure 2 shows examples of TAKUMI's north-bank and south-bank neutron diffractograms, acquired for 90 min at room temperature after isothermal annealing. Here, these diffractograms were comparably plotted according to the lattice plane spacing (d) and the momentum transfer ($Q = 2\pi/d$), respectively. In the north-bank diffractogram, the ferrite-110 peak was much higher than that in the south-bank diffractogram (about 17,000 counts vs. 10,900 counts), while the ferrite-200 peaks had almost the same intensity (about 710 counts vs. 760 counts); on the other hand, the austenite-200 peak in the north-bank diffractogram was higher than that in the south-bank diffractogram (about 3450 counts vs. 1910 counts), while their austenite-111 peaks had almost the same intensity (about 5600 counts vs. 5500 counts). These two different neutron diffractograms of the investigated steel specimens from the north bank and the south bank (corresponding to the RD and the TD, respectively) revealed that a strong texture indeed exists in the steel specimens after isothermal annealing.

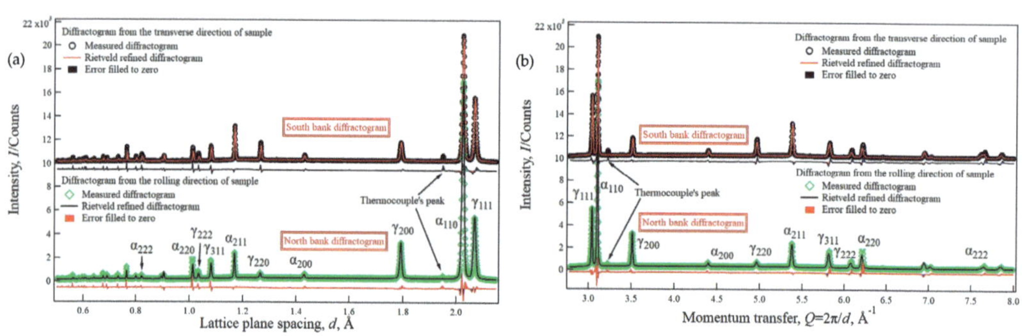

Figure 2. North-bank and south-bank time-of-flight neutron diffractograms and their Rietveld-refined results obtained after the TAKUMI in situ neutron diffraction experiment: (**a**) lattice plane spacing, where the low (hkl)-index lattice planes with large lattice spacing are usually paid more attention during rapid preferred orientation/texture measurement in the real space; (**b**) momentum transfer, usually involved in the calculation of the orientation distribution function in the reciprocal space.

Using the GSAS Rietveld refinement considering the orthorhombic sample symmetry, the residual of least-squares refinement R_p and the weighted residual of least-squares refinement R_{wp} were 0.075/0.079 and 0.116/0.132 for the north-/south-bank neutron profiles, respectively, revealing that the full Rietveld refinement was satisfactory. The mass fraction of austenite was $37.0 \pm 0.4\%$ from the north-bank diffractogram, and it was about $39.3 \pm 0.8\%$ from the south-bank diffractogram, providing a good reference for the transformation evolution process, although the weak neutron intensities of the high-index diffraction peaks easily led to a large refinement error. The lattice parameter of ferrite was $a = 2.86309 \pm 0.00004$ Å from the north-bank diffractogram and $a = 2.86297 \pm 0.00004$ Å from the south-bank diffractogram. The lattice parameter of austenite was $a = 3.58158 \pm 0.00008$ Å from the north-bank diffractogram and $a = 3.58067 \pm 0.00008$ Å from the south-bank diffractogram. The relative errors in the lattice parameters of ferrite and austenite were about 4.2×10^{-5} and 2.5×10^{-4}, respectively, resulting from the slight error in position and the possible residual strain.

Figure 3 shows the inverse pole figures extracted from the north-bank and south-bank neutron diffractogram refinement. After the GSAS Rietveld analysis including spherical harmonic (ODF) preferred orientations, we can obtain the fitting process list, including the preferred orientation parameter known as "Prfo", which is the relative intensity ratio of each hkl peak after considering the multiplicity and the structure factor of each hkl reflection. We constructed a data array including the plane coordinate (x_{hkl}, y_{hkl}) of each inverse pole figure orientation hkl and its relative intensity ratio ($\mathrm{Prfo}_{\mathrm{hkl}}$), and then we can draw a contour plot using Igor Pro 8.0/OriginPro 9.6 or other plotting software for all *hkl*-index orientations to construct Figure 3, marking all the orientation indices in advance in the figure. It should be noted that more than eight of such *hkl* peaks are needed for each phase, i.e., the time-of-flight neutron diffractogram should have high neutron counts, and its lattice plane spacing range should be wide enough.

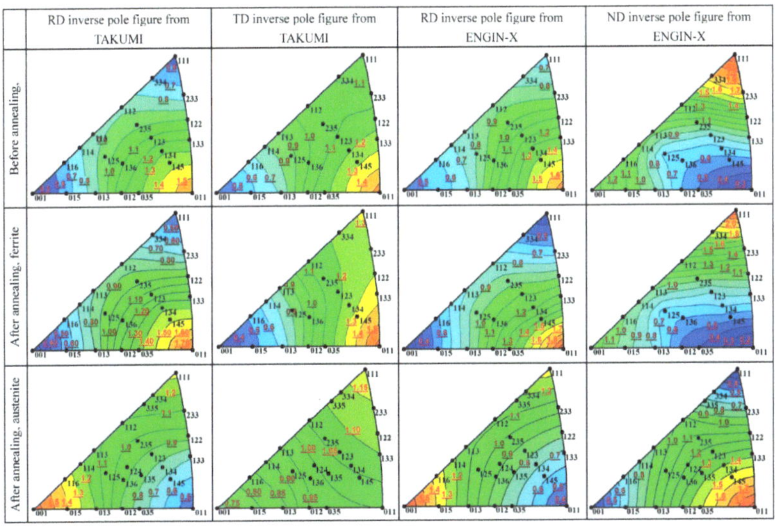

Figure 3. Inverse pole figures before and after the in situ neutron diffraction experiments, extracted from the north-bank and south-bank time-of-flight neutron diffractograms of TAKUMI and ENGIN-X: before annealing, cold-rolled martensite; after annealing, statically recrystallized ferrite and precipitated austenite. The red underlined text in the inverse pole figures is employed for the orientation distribution intensity to mark the contour lines, where 1.0 is for random, >1.0 for higher orientation distribution intensity and <1.0 for lower orientation distribution intensity.

From the RD inverse pole figures from TAKUMI, it was clear that the crystallographic orientations of martensite/ferrite were relatively concentrated from the orientations [100]//RD and [111]//RD to the orientation [011]//RD during the isothermal annealing. For the ENGIN-X neutron diffraction with 90° TD → ND rotation of the investigated specimen relative to the TAKUMI sample orientation, the obtained RD inverse pole figures of martensite/ferrite showed good consistency in the crystallographic orientation distribution, and their texture evolution trend was almost same. Moreover, good consistency was also found in the RD inverse pole figures of precipitated austenite during the isothermal annealing. This suggests that reliable preferred orientation characteristics were obtained from the neutron diffractograms even though the samples were in different orientation setups for the different neutron diffractometers. For the TD inverse pole figures from TAKUMI, it was found that the crystallographic orientations of ferrite concentrated from the orientation [010]//TD to the orientations [101]//TD and [111]//TD; for the ND inverse pole figures from ENGIN-X, it was found that the crystallographic orientations of the ferrite grains became relatively sharper at the orientation [111]//ND due to the static recrystallization. The precipitated austenite showed a principal preferred orientation with [100]//RD, [111]//TD and [110]//ND.

Figure 4 shows the RD, TD, and ND triaxial inverse pole figures calculated from the complete pole figures measured on MUSASI-L. It was evident that for the martensite/ferrite, the orientations [100]//RD and [111]//RD concentrated to the orientation [110]//RD during the isothermal annealing, while the orientations [101]//TD and [111]//ND became much sharper during the isothermal annealing. The precipitated austenite showed a principal preferred orientation with [100]//RD, [111] //TD, and ~[110]//ND. The similar trends of principal preferred orientation distribution evolution suggested that the GSAS full Rietveld refinement together with a spherical harmonic function is very valuable to semi-quantitatively monitor the in situ microstructure/texture evaluation under some extreme environments.

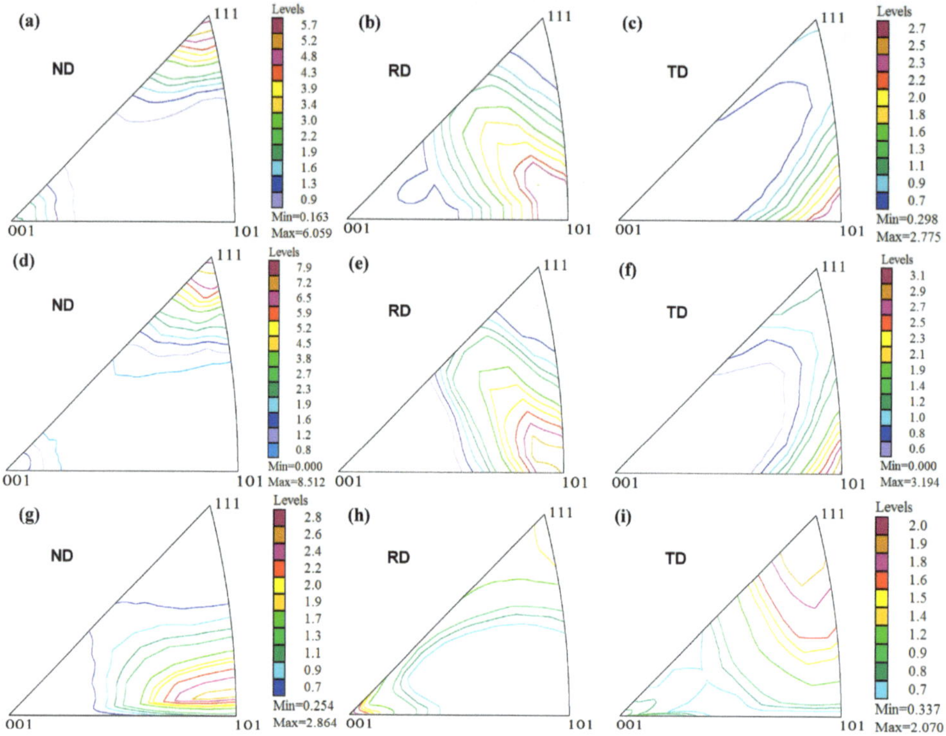

Figure 4. Inverse pole figures (**a–c**) before and (**d–i**) after the in situ neutron experiments, calculated from the complete pole figures measured on MUSASI-L: (**a**) ND inverse pole figure of cold-rolled martensite; (**b**) RD inverse pole figure of cold-rolled martensite; (**c**) TD inverse pole figure of cold-rolled martensite; (**d**) ND inverse pole figure of statically recrystallized ferrite; (**e**) RD inverse pole figure of statically recrystallized ferrite; (**f**) TD inverse pole figure of statically recrystallized ferrite; (**g**) ND inverse pole figure of precipitated austenite; (**h**) RD inverse pole figure of precipitated austenite; (**i**) TD inverse pole figure of precipitated austenite.

On the other hand, it should be mentioned that there was a clear difference in the absolute orientation distribution intensities of the inverse pole figures obtained from the bulk texture measurements and the time-of-flight neutron diffraction full-diffractogram refinements. The most important reason for this can be attributed to the different analysis methods: the MUSASI-L result using the direct discrete LaboTex ADC method [25] avoids any truncation errors of the expansion series of the spherical harmonic method (note: the background noise information comparable with the real diffraction intensity information of textured materials during a rapid texture measurement is possibly overcalculated, resulting in a low symmetry of the recalculated pole figures), while for the time-of-flight neutron diffraction profile refinement the lower expansion series of the generalized spherical harmonic function ($L_{max} = 8$) enables a full-diffractogram refinement with a relatively smooth preferred orientation evaluation from only one time-of-flight diffractogram, at the expense of larger truncation errors [19,20]. If the multiple time-of-flight neutron diffractograms with wide stereographic angle coverage are employed simultaneously to provide more fruitful orientation information, the available higher maximum expansion series L_{max} may enable the full-diffractogram refinement to achieve a high-precision preferred orientation evaluation, including the calculation of pole figures and orientation distribution function.

Moreover, the above difference is also related to the lower stereographic angle resolution of the north-bank and south-bank neutron detectors for each diffractogram, averaged by the large stereographic angle coverage of TAKUMI (and ENGIN-X) in the case of no stereographic angle division of the neutron detector panels. When using the high stereographic angle resolution method involved in our recent TAKUMI time-of-flight neutron diffraction bulk texture measurement technique [26], a better reliability of principal preferred orientation evolution can be achieved for most in situ neutron diffraction materials experiments. Here, during the angle-dispersive neutron diffraction measurements, the neutron diffraction peaks were detected in a small stereographic angle coverage of the single-tube detector (less than $2^\circ \times 2^\circ$ in $\Delta\chi$ and $\Delta\varphi$, defined by the 20 mm $\times$ 20 mm beam slits at about 600 mm from the sample center) at each stereographic angle grid of $5^\circ \times 5^\circ$.

Figure 5 shows the inverse pole figures obtained from the MAUD Rietveld texture analysis before (a,b) and after (c,d) isothermal annealing during the HIPPO in situ neutron diffraction experiment. It can be seen here that there was 1.4 $\pm$ 0.7 mass% austenite retained in the cold-rolled martensitic matrix, and its principal preferred orientation characteristics were successfully extracted, confirming that the phase fraction analysis method of retained austenite through the combined Rietveld texture analysis of a textured steel [27] is highly reliable. However, 1~2 full-range time-of-flight neutron diffractograms of textured steels containing 1.4 $\pm$ 0.7 mass% retained austenite with a certain principal preferred orientation cannot provide enough reliable diffraction peak intensity information to plot the inverse pole figure of a tiny amount of austenite. After the isothermal annealing, the recrystallized ferrite and the precipitated austenite had a good mass fraction balance, allowing them to be easily evaluated using a full-range time-of-flight neutron diffractogram. Comparing the corresponding inverse pole figures in Figures 3 and 5, a similar orientation distribution can be found in the cold-rolled martensite before isothermal annealing, the recrystallized ferrite, and the precipitated austenite after isothermal annealing. Such similar orientation distribution characteristics among these inverse pole figures suggest that the principal crystallographic orientation evaluation method based on time-of-flight neutron diffractograms can be reasonably applied to in situ crystallographic structure evolution studies of bulk orthorhombic materials and spatially resolved principal preferred orientation mappings of large engineering structure parts.

From the HIPPO combined Rietveld texture analysis, the lattice parameter of ferrite was found to be a = 2.866 $\pm$ 0.002 Å, and the lattice parameter of austenite was a = 3.585 $\pm$ 0.003 Å, while the phase fraction of precipitated austenite was about 36.8 $\pm$ 0.8 mass%. Here, the larger error of the lattice parameters for austenite and ferrite was mostly related to the combined usage of $2\theta = 40^\circ$ detector panels with a low instrumental resolution of $\Delta d/d$ = 1.8%; for HIPPO, $2\theta = 145^\circ$ detector panels with a high instrumental resolution of $\Delta d/d$ = 0.37% are usually employed to evaluate the lattice parameters of powder samples [23].

It should be noted that the phase fraction of retained austenite in the cold-rolled 17Ni-0.2C steel before 823 K isothermal annealing was not well analyzed in the GSAS fitting of single TAKUMI/ENGIN-X diffractograms (Figure 6), while the phase fraction of austenite from HIPPO as 1.4 $\pm$ 0.7 mass%. This inconsistency is mainly related to the 120 neutron diffractograms from various sample orientations employed together in the HIPPO texture analysis, where some diffractograms form specific orientations that may provide distinguishable austenite peaks for the phase fraction analysis. If we carry out a high-stereographic-resolution texture measurement with TAKUMI and/or ENGIN-X [26,27], we may also obtain a reliable phase fraction of retained austenite. For MUSASI-L, the single-tube neutron detector does not have a high enough angle accuracy to detect the tiny second phase of austenite before in situ neutron diffraction.

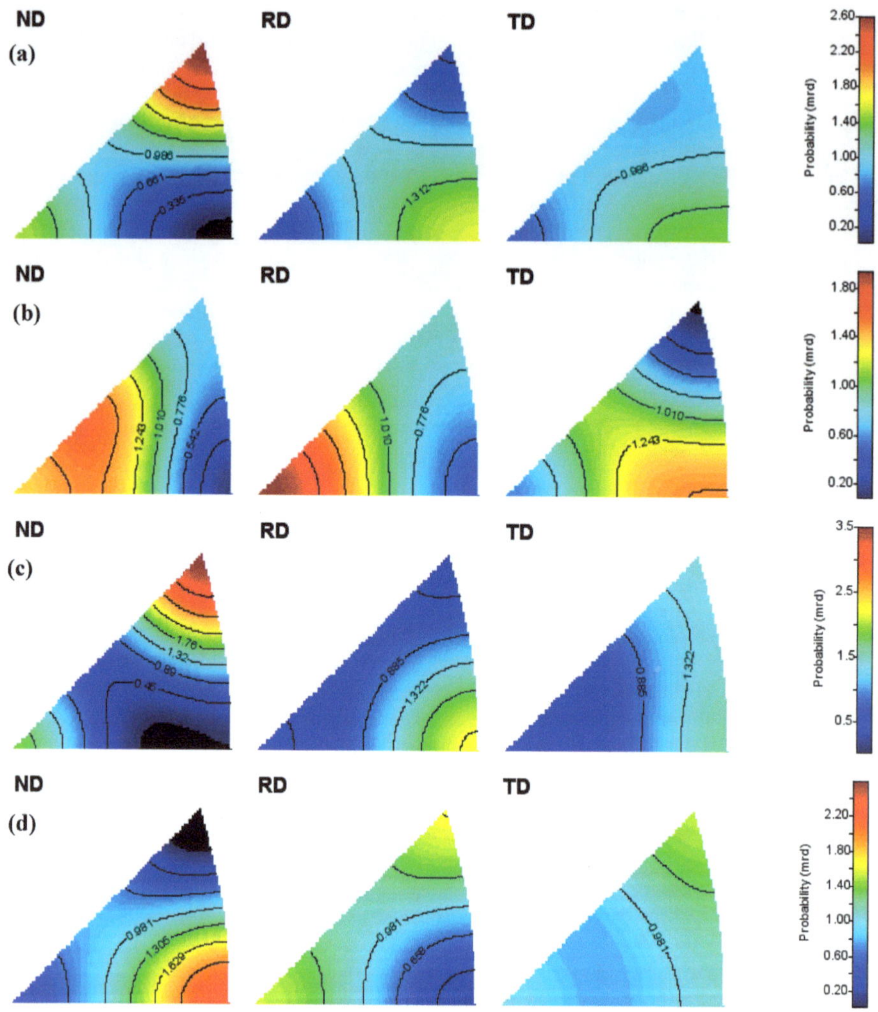

Figure 5. Inverse pole figures (**a**,**b**) before and (**c**,**d**) after the HIPPO in situ neutron experiment, obtained from MAUD Rietveld texture analysis: (**a**) cold-rolled martensite and (**b**) retained austenite (phase fraction: 1.4 ± 0.7 mass%); (**c**) recrystallized ferrite and (**d**) precipitated austenite (36.8 ± 0.8 mass%).

Figure 6 primarily reveals the competitive microstructural evolution process to obtain an ultrafine-grained multiphase microstructure involved in (a) the static recrystallization of cold-rolled martensite (with broadened and flattened peaks) to obtain newly recrystallized ferrite grains with sharp and strong ferrite-110 peaks and (b) the static precipitation of austenite grains with sharp and strong austenite-111 peaks.

If the main research purpose is related to the lattice structural changes and stress/strain evolution during an in situ time-of-flight neutron diffraction experiment, the engineering materials neutron diffractometers with high stereographic angles and spatial resolutions [26,27] may be recommended; if the main research purpose is related to the phase fraction and the textures, the powder diffractometers with high stereographic angle coverage [8,23,28–30] are3 high-throughput candidates. The complementary use of two kinds of time-of-flight neutron diffractometer may provide us with more fruitful information for a deeper understanding of the microstructure and texture evolutions of advanced multiphase materials in some extreme environments.

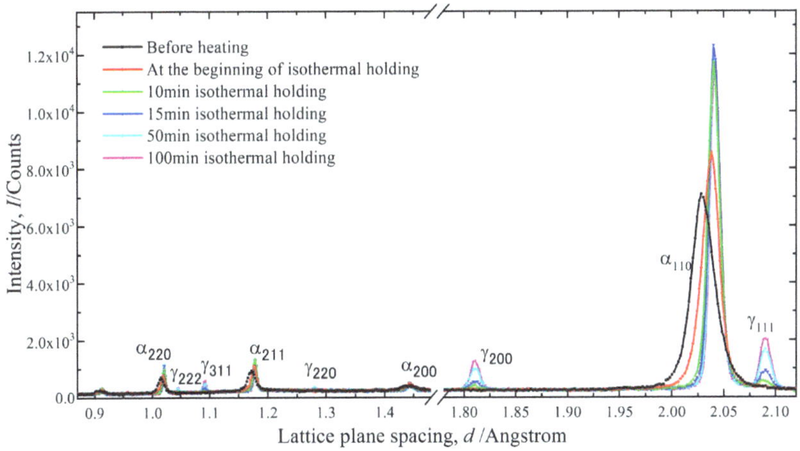

Figure 6. Change in the ENGIN-X time-of-flight neutron diffractograms obtained along the rolling direction of 17Ni-0.2C steel before and during 823 K isothermal holding.

5. Conclusions

Based on the GSAS full Rietveld refinement of the TAKUMI north-bank and south-bank neutron diffractograms, the principal preferred orientation analysis was carried out, and the extracted inverse pole figures were compared with those obtained from ENGIN-X neutron diffraction. The consistent RD inverse pole figures before and after the isothermal annealing suggested that the principal preferred orientation evaluation of bulk materials can be realized through 90° TD → ND rotation of the investigated specimen during two neutron diffraction experiments. The same texture evolution trend was confirmed through the bulk texture measurements using the MUSASI-L angle-dispersive neutron diffraction. The similar orientation distribution characteristics suggested that the above inverse pole figure analysis method based on the time-of-flight neutron diffraction full-diffractogram refinement may be applied to monitor the changes in the principal preferred orientation evolution of bulk orthorhombic materials during various in situ neutron diffraction experiments.

Author Contributions: Conceptualization, P.X. and Y.T.; sample preparation, P.X. and Y.T., experimental methodology, P.X.; neutron diffraction experiments, P.X., S.Z., S.H. and S.C.V.; experimental data conversion, P.X., S.Z., S.H. and S.C.V.; neutron diffraction analysis, P.X.; investigation, P.X. and Y.T.; neutron facility resources, P.X., S.Z., S.H. and S.C.V.; writing—original draft preparation, P.X.; writing—review and editing, P.X., S.Z., S.H., Y.T. and S.C.V.; supervision, P.X.; project administration, P.X.; funding acquisition, P.X. and S.H. All authors have read and agreed to the published version of the manuscript.

Funding: This research project was financially supported by the Japan Society for the Promotion of Science (Grant No. 21860090, JP17K06852) and by the MEXT Program: Data Creation and Utilization Type Materials Research and Development (JPMXP1122684766).

Data Availability Statement: Data will be made available on request.

Acknowledgments: The neutron experiment at the Materials and Life Science Experimental Facility of J-PARC was performed under a user program (Proposal No. 2009B0006); the neutron experiment at ISIS was performed under a user program (Proposal No. RB1110337); the neutron experiment at LANL was performed under a user program (Proposal No. 20102030).

Conflicts of Interest: The authors declare no conflicts of interest.

References

1. Raabe, D. Textures of strip cast and hot-rolled ferritic and austenitic stainless steel. *Mater. Sci. Technol.* **1995**, *11*, 461–468. [CrossRef]
2. Liu, M.X.; Song, C.J.; Cui, Z.S. Crystallographic texture evolution and martensite transformation in the strain hardening process of a ferrite-based low density steel. *J. Mater. Sci. Technol.* **2021**, *78*, 247–259. [CrossRef]
3. Bunge, H.J. Advantages of neutron diffraction in texture analysis. *Textures Microstruct.* **1989**, *10*, 265–307. [CrossRef]
4. Wenk, H.R.; Huensche, I.; Kestens, L. In situ observation of texture changes during phase transformations in ultra-low-carbon steel. *Metall. Mater. Trans. A* **2007**, *38*, 261–267. [CrossRef]
5. Gurao, N.P.; Ashkar, A.A.; Suwas, S. Study of texture evolution in metastable β-Ti alloy as a function of strain path and its effect on α transformation texture. *Mater. Sci. Eng. A* **2009**, *504*, 24–35. [CrossRef]
6. Brokmeier, H.G.; Gan, W.M.; Randau, C.; Völler, M.; Rebelo-Kornmeier, J.; Hofmann, M. Texture analysis at neutron diffractometer STRESS-SPEC. *Nucl. Instrum. Methods Phys. Res. A* **2011**, *642*, 87–92. [CrossRef]
7. Wei, Z.G.; Yang, P.; Gu, X.F.; Onuki, Y.; Sato, S. Transformation textures in pure titanium: Texture memory vs surface effect. *Mater. Charact.* **2020**, *164*, 110359. [CrossRef]
8. Xu, P.G.; Liss, K.D. Multiple wavelengths texture measurement using angle dispersive neutron diffraction at WOMBAT. *Quantum Beam Sci.* **2021**, *5*, 11. [CrossRef]
9. Daymond, M.R.; Bourke, M.A.M.; Von Dreele, R.B.; Clausen, B.; Lorentzen, T. Use of Rietveld refinement for elastic macrostrain determination and for evaluation of plastic strain history from diffraction spectra. *J. Appl. Phys.* **1997**, *82*, 1554–1562. [CrossRef]
10. Sun, Y.N.; Choo, H.; Liaw, P.K.; Lu, Y.L.; Yang, B.; Brown, D.W.; Bourke, M.A.M. Neutron diffraction studies on lattice strain evolution around a crack-tip during tensile loading and unloading cycles. *Scr. Mater.* **2005**, *53*, 971–975. [CrossRef]
11. Ma, D.; Stoica, A.D.; An, K.; Yang, L.; Bei, H.; Mills, R.A.; Skorpenske, H.; Wang, X.L. Texture Evolution and Phase Transformation in Titanium Investigated by *In Situ* Neutron Diffraction. *Metall. Mater. Trans. A* **2011**, *42*, 1444–1448. [CrossRef]
12. Matthies, S.; Lutterotti, L.; Wenk, H.R. Advances in texture analysis from diffraction spectra. *J. Appl. Crystallogr.* **1997**, *30*, 31–42. [CrossRef]
13. Wang, X.L. *Conceptual Design of the SNS Engineering Diffractometer*; SNS Report No. IS-1.1.8.2-6035-RE-A-00; Oak Ridge National Laboratory: Oak Ridge, TN, USA, 2000.
14. Kockelmann, W.; Chapon, L.C.; Radaelli, P.G. Neutron texture analysis on GEM at ISIS. *Phys. B* **2006**, *385–386*, 639–643. [CrossRef]
15. Ito, T.; Nakatani, T.; Harjo, S.; Arima, H.; Abe, J.; Aizawa, K.; Moriai, A. Application software development for the engineering materials diffractometer, TAKUMI. *Mater. Sci. Forum* **2010**, *652*, 238–242. [CrossRef]
16. Oishi, R.; Yonemura, M.; Nishimaki, Y.; Torii, S.; Hoshikawa, A.; Ishigaki, T.; Morishima, T.; Mori, K.; Kamiyama, T. Rietveld analysis software for J-PARC. *Nucl. Instr. Methods Phys. Res. A* **2009**, *600*, 94–96. [CrossRef]
17. Dollase, W.A. Correction of intensities for preferred orientation in powder diffractometry—Application of the March model. *J. Appl. Crystallogr.* **1986**, *19*, 267–272. [CrossRef]
18. Von Dreele, R.B. Quantitative texture analysis by Rietveld refinement. *J. Appl. Crystallogr.* **1997**, *30*, 517–525. [CrossRef]
19. Bunge, H.J. *Texture Analysis in Materials Science*; Butterworths: London, UK, 1982.
20. Larson, A.C.; Von Dreele, R.B. *General Structure Analysis System (GSAS)*; Los Alamos National Laboratory Report LAUR 86-748; The Regents of the University of California: Oakland, CA, USA, 2004.
21. Harjo, S.; Moriai, A.; Torii, S.; Suzuki, H.; Suzuya, K.; Morii, Y.; Arai, M.; Tomota, Y.; Akita, K.; Akiniwa, Y. Design of engineering diffractometer at J-PARC. *Mater. Sci. Forum* **2006**, *524–525*, 199–204. [CrossRef]
22. Haynes, R.; Paradowska, A.M.; Chowdhury, M.A.H.; Goodway, C.M.; Done, R.; Kirichek, O.; Oliver, E.C. An inert-gas furnace for neutron scattering measurements of internal stresses in engineering materials. *Meas. Sci. Technol.* **2012**, *23*, 047002. [CrossRef]
23. Wenk, H.R.; Lutterotti, L.; Vogel, S.C. Texture analysis with the new HIPPO TOF diffractometer. *Nucl. Instrum. Methods Phys. Res. A* **2003**, *515*, 575–588. [CrossRef]
24. Lutterotti, L.; Matthies, S.; Wenk, H.R.; Schultz, A.S.; Richardson, J.W.J. Combined texture and structure analysis of deformed limestone from time-of-flight neutron diffraction spectra. *J. Appl. Phys.* **1997**, *81*, 594–600. [CrossRef]
25. Pawlik, K.; Pospiech, J.; Lücke, K. The ODF approximation from pole figures with the aid of the ADC method. *Textures Microstruct.* **1991**, *14–18*, 25–30. [CrossRef]
26. Xu, P.G.; Harjo, S.; Ojima, M.; Suzuki, H.; Ito, T.; Gong, W.; Vogel, S.C.; Inoue, J.; Tomota, Y.; Aizawa, K.; et al. High stereographic resolution texture and residual stress evaluation using time-of-flight neutron diffraction. *J. Appl. Crystallogr.* **2018**, *51*, 746–760. [CrossRef]
27. Xu, P.G.; Tomota, Y.; Arakaki, Y.; Harjo, S.; Sueyoshi, H. Evaluation of austenite volume fraction in TRIP steel sheets using neutron diffraction. *Mater. Charact.* **2017**, *127*, 104–110. [CrossRef]
28. Onuki, Y.; Hoshikawa, A.; Sato, S.; Ishigaki, T.; Tomida, T. Quantitative phase fraction analysis of steel combined with texture analysis using time-of-flight neutron diffraction. *J. Mater. Sci.* **2017**, *52*, 11643–11658. [CrossRef]

29. Mori, M.; Yamanaka, K.; Onuki, Y.; Sato, S.; Chiba, A. Analysis of hierarchical microstructural evolution in electron beam powder bed fusion Ti–6Al–4V alloys via time-of-flight neutron diffraction. *Addit. Manuf. Let.* **2022**, *3*, 100053. [CrossRef]
30. Li, Q.; Onuki, Y.; Ahadi, A.; Sun, Q.P. Large tunable thermal expansion in ferroelastic alloys by stress. *Acta Mater.* **2022**, *240*, 118350. [CrossRef]

MDPI AG
Grosspeteranlage 5
4052 Basel
Switzerland
Tel.: +41 61 683 77 34

Quantum Beam Science Editorial Office
E-mail: qubs@mdpi.com
www.mdpi.com/journal/qubs

MDPI